BLOCKBUSTER DRUGS

BLOCKBUSTER DRUGS

The Rise and Decline of the Pharmaceutical Industry

JIE JACK LI

OXFORD
UNIVERSITY PRESS

OXFORD
UNIVERSITY PRESS

Oxford University Press is a department of the University of Oxford.
It furthers the University's objective of excellence in research, scholarship,
and education by publishing worldwide.

Oxford   New York
Auckland   Cape Town   Dar es Salaam   Hong Kong   Karachi
Kuala Lumpur   Madrid   Melbourne   Mexico City   Nairobi
New Delhi   Shanghai   Taipei   Toronto

With offices in
Argentina   Austria   Brazil   Chile   Czech Republic   France   Greece
Guatemala   Hungary   Italy   Japan   Poland   Portugal   Singapore
South Korea   Switzerland   Thailand   Turkey   Ukraine   Vietnam

Oxford is a registered trademark of Oxford University Press
in the UK and certain other countries.

Published in the United States of America by
Oxford University Press
198 Madison Avenue, New York, NY 10016

© Oxford University Press 2014

[Library of Congress Cataloging-in-Publication Data
Li, Jie Jack, author.
Blockbuster drugs : the rise and decline of the pharmaceutical industry / Jie Jack Li.
p. ; cm.
Includes bibliographical references and index.
Summary: "This book uses the cases of several landmark drugs to discuss the history of
the pharmaceutical industry, and discusses what could be next"--Provided by publisher.
ISBN 978-0-19-973768-0 (hardback : alk. paper)
I. Title.
[DNLM: 1. Drug Industry—history—United States.   2. Drug Industry—
trends—United States.   3. Drug Therapy—history—United States.   4. History,
20th Century—United States.   5. History, 21st Century—United
States.   6. Pharmaceutical Preparations—history—United States. QV 11 AA1]
RS100
338.4'76151—dc23
2013017116

9 8 7 6 5 4 3 2 1
Printed in the United States of America
on acid-free paper

To Professor Rick L. Danheiser

Contents

Preface

British statesman and philosopher Edmund Burke (1729–1797) once said, "Those who don't know history are destined to repeat it." For science in general and drug discovery in particular, history is probably the best teacher from which we can learn invaluable lessons. While tremendous resources, both matériel and human, have been invested in drug discovery, the majority of these investments have not been as fruitful as we would hope, especially during the last decade. Therefore, we could learn *what not to do* from failures. On the other hand, there have been a number of spectacular successes in drug discovery, as represented by blockbuster drugs, which are the focus of this book. We could learn *what to do* from the successes.

The second half of the 20th century was the golden age of drug discovery. Advances in medicine saved lives and/or improved the quality of lives. Meanwhile, the pharmaceutical industry flourished. In this book, I have recounted the discovery and development of several classes of blockbuster drugs in the fields of ulcer drugs, allergy medicines, blood thinners, and analgesics. One class of blockbuster drugs not covered here is statins;

almost every single one of them was a mega blockbuster. My previous book, *Triumph of the Heart: The Story of Statins*, thoroughly chronicled that story.

I hope both scientists and the general public will enjoy reading this book and appreciate how far we have come in terms of progress in medicines. I could not be happier if the reader is inspired by learning the fascinating history of blockbuster drugs.

I am grateful for permissions to use the postage stamps from respective postal authorities, which still retain the copyrights for those stamps. The chemical structures for the drugs that are printed in **bold** are compiled in the appendix.

As always, I welcome your critiques. You can e-mail your comments directly to me at lijiejackli@gmail.com.

Jack Li
April 1, 2013
Princeton, New Jersey

BLOCKBUSTER DRUGS

1

Before the Age of Blockbuster Drugs

We try to remember that medicines are for the patient. We try never to forget that medicine is for the people. It is not for profit. The profits follow and if we've remembered that, they have never failed to appear. The better we have remembered it, the larger they have been.

GEORGE W. MERCK, *former president of Merck & Co. (1894–1957)*

Blockbuster drugs are drugs with annual sales over $1 billion these days. As a testimony of changing times, a blockbuster drug was defined as a drug with more than $500 million in annual sales just a decade ago! While these blockbuster drugs save millions of lives and improve the quality of life for millions of others, pharmaceutical companies make considerable profit. In turn, drug makers then spend a tremendous amount of money on research and development of new blockbuster drugs, looking for new ones that will sustain the "life cycle" for the health of both patients (physical and mental) and the drug companies themselves (financial).

Pharmaceutical companies are sometimes known as "merchants of life." Indeed, their products affect people's lives in many positive ways. But make no mistake; the drug industry is a for-profit entity, responsible to its shareholders. It must make a profit to survive. This is the contradiction of the pharmaceutical industry.

However, things were not always like this until the last few decades.

The first antihistamine (the substance that counteracts the effects of histamine; see chapter 4 for more details) was discovered by French pharmacologist Daniel Bovet in 1937.[1] Between 1937 and 1941, Bovet conducted more than 3,000 experiments to find the chemical formulas upon which most of the antihistamines now prescribed are based. Antihistamines are effective in treating allergic reactions. His discovery led to development of the first antihistamine drug, **diphenhydramine** (Antergan), for treating allergies in 1942, but it did not reach the market because of toxicity issues. In 1944, another one of Bovet's discoveries, **pyrilamine** (Neoantergan), was produced as a drug. He did not patent it and did not make a penny out of his important discovery. Not all was lost, however; Bovet won the Nobel Prize in Physiology or Medicine in 1957.

General Robert Wood Johnson (1845–1910), one of the three brothers who founded Johnson & Johnson, wrote a credo that codified the company's socially responsible approach to conducting business. The credo stated that the company's responsibility is first to the people who use its products and services, second to its employees, third to the community and environment, and fourth to its shareholders.[2] In good times, it is easy to be magnanimous, but during tough times, profit seems to trump other values, especially when many drug companies are now run by businesspeople. Some may feel that profit has become the drug companies' top priority.

In 1943, Selman Abraham Waksman, a professor of soil microbiology at Rutgers University, and his student Albert Schatz isolated an antibiotic that was very effective at killing Gram-negative bacteria, which were not killed by penicillin. Waksman christened it **streptomycin**. With assistance from the Mayo Clinic in animal testing and clinical trials, and Merck for large-scale production, streptomycin was proven to be both safe and effective at treating tuberculosis. It was one of the first effective drugs against tuberculosis, an infectious disease that has plagued humanity for millennia. In 1938, Waksman and Merck's then-CEO George W. Merck (1894–1957) struck a deal resulting in the establishment of a Merck fellowship in fermentation studies in Waksman's laboratory. Under the agreement, Waksman received chemistry support from Merck and was able to use the extensive facilities that Merck could offer for streptomycin production. In return, Merck received the patents on any process that Waksman developed.[3] Rutgers University licensed streptomycin to Merck. In an extraordinarily benevolent gesture for a drug firm, George W. Merck, at the request of Waksman, signed away the exclusive right for developing streptomycin. This meant that any drug company on earth could make and sell streptomycin. Merck's magnanimous act greatly enhanced his company's reputation not only for its science, but also for its humanitarian fame. As a result of competition among the many drug firms, the price of streptomycin quickly became very affordable.

While giving a speech at the Medical College of Virginia at Richmond in 1950, George W. Merck immortalized his company's values with the words that began this chapter and which would resonate for decades to come.[3]

Years later in 1987, Merck's then-CEO P. Roy Vagelos made a similarly generous decision that Merck would donate Mectizan free to anyone in the world who needed it. **Ivermectin** (Mectizan) is an antiparasite drug that is very effective at preventing river blindness, a devastating disease transmitted by the black fly that has plagued sub-Saharan Africa and some countries in Latin America. Merck donated in excess of 2.5 billion tablets of Mectizan at a total cost of more than $3.75 billion over the years.[3] During Vagelos's reign, Merck was not only the number-one pharmaceutical company in the world; it was also the most admired company for its top-notch science, corporate citizenship, and decency. It was many graduate students' dream, among others, to work for Merck. Scientists at Merck were, deservedly, very proud, and even arrogant according to some. Today, the perception of the drug industry is decidedly less rosy. Even Merck, the golden standard of corporate citizenship, earned a dubious reputation from the repercussions of the withdrawal of Vioxx from the market in 2004 when it was found to have cardio toxicity. The painkiller Vioxx was a bona fide blockbuster drug, making $2.5 billion in 2003. However, the problems with Vioxx cost Merck more than $4.5 billion in legal expenses.

Meanwhile, Pfizer became the largest drug company in the world by gobbling up Warner-Lambert in 2000, Pharmacia in 2003, and Wyeth in 2007. In addition to several spectacular failures of big drugs such as **torcetrapib** and Exubera (the Titanic and Hindenburg disasters were frequently cited as analogies), Pfizer's reputation, along with the drug industry as a whole, had been tarnished as well. "What has broken down is the perception of our values," according to former Pfizer CEO Jeffrey Kindler[4] in 2009, "We were seen as greedy, arrogant, not transparent."[5]

The pharmaceutical industry has a long way to go to restore its reputation. The public has largely forgotten that drug companies used to be among the most admired businesses for a long time during the last century. Now, I give you a look at five classes of blockbuster drugs, with a strong sense of nostalgia for the golden age of the pharmaceutical industry, but also with hope for the industry's renaissance in the future by learning invaluable lessons from past successes and failures.

Beginning of an Era

The First Blockbuster Drug, Tagamet

*The most fruitful basis for the discovery of a new drug is to
start from an old drug.*

JAMES W. BLACK, *1988 Nobel Laureate (1924–2010)*

Tagamet emerged as the first blockbuster drug when its sales
exceeded $1 billion in 1986, three years after its introduction to
the market. An anti–peptic ulcer drug, Tagamet was discovered by
James W. Black and his colleagues at Smith Kline & French's (SK&F)
British subsidiary in Welwyn Garden City. Before Tagamet, SK&F
was a little-known U.S. drug firm in Philadelphia. After Tagamet,
SK&F became one of the largest pharmaceutical companies in the
world. The history of Tagamet is one of the most extraordinary in
the annals of medicine. It is a saga of a drug that almost escaped
detection because the research efforts that began in 1964 did not
seem to produce results within the first *11 years!*

§2.1. James Black and the Discovery of Tagamet

Smith Kline & French

Smith Kline started as a humble drug store in Philadelphia
in 1830. During the American Civil War, Smith Kline
was founded as a small apothecary by two physicians,

John K. Smith and John Gilbert on North Second Street.[1] Not only was Philadelphia the birthplace of the United States of America, it was also the cradle of American pharmacy. Wyeth, McNeil, Rorer, and Warner-Lambert all trace their origins to small drug stores established there during the Civil War. In the 1880s, Mahlon N. Kline led the company into research and manufacturing of its own products. In 1891, it absorbed French, Richards & Co. founded by Harry B. French, creating Smith Kline & French. After its establishment, the company slowly expanded its inventory. By the 1920s, it had some 15,000 products ranging from aspirin to liniment. Their Eskay's Albumenized Food was highly popular as a digestible food for infants and the disabled. Later, the company did very well with Eskay's Tablets for Seasickness. Its specialty, Eskay's Neurophosphates, a nerve tonic, soothed millions of people at home and abroad. In 1929, Smith Kline & French Laboratories was created to devote itself solely to research and development (R&D).

During the Great Depression year of 1936, the company stepped up its efforts in R&D (in a recent contrast, many pharmaceutical companies stepped down their R&D investments during the last recession of 2008). In addition to finding a few new tranquilizers, SK&F was also "fortunate" enough to discover **amphetamine**, a drug that was initially intended to "pep up" tired individuals but has been widely abused since then. In 1955, the company introduced to America the antipsychotic **chlorpromazine** (Thorazine), synthesized by Paul Charpentier in the French drug firm Rhône-Poulenc in 1950. Its antipsychotic properties were discovered by French physician Henri-Marie Laborit. One year after Thorazine's introduction, U.S. mental hospitals discharged more patients than they admitted for the first time in history. SK&F also pioneered some innovative marketing strategies. They were the first to regularly send doctors samples of new drugs through the mail.

In early 1959, SK&F opened and expanded a small research outfit in Welwyn Garden City north of London in Hertfordshire. The rationale was that the British researchers might have different perspectives on approaching drug discovery than their U.S. counterparts. In this case, SK&F's small investment in the Welwyn Research Institute paid handsome dividends with Tagamet. SK&F was a small, quiet drug firm until 1976 when it launched **cimetidine** (trade name Tagamet), the first effective drug to treat peptic ulcers (holes in the lining of the stomach and the duodenum, the upper part of the small intestine). Tagamet catapulted SK&F into the top-player level in the pharmaceutical industry. Much credit is owed to James W. Black, who worked as a senior pharmacologist at Welwyn Research Institute from 1964 to 1972, for the discovery of Tagamet

James Black and Beta-Blockers

The fourth of five sons in a working-class family, James Whyte Black was born on June 14, 1924, in Uddington, Scotland. His father began working as a coal miner at age 12, but he used night school to advance his career, rising to become a mining engineer.[2] Young James was strongly influenced by his father's work ethics. After finishing his undergraduate degree from St. Andrews University in 1946, James Black did physiology studies there for a year. He then moved to Singapore to teach for three years to pay off his debts accumulated from his medical studies. When he came back to England, the job market was so tight that he became really desperate until he was asked to start a new Physiology Department at the recently "nationalized" University of Glasgow Veterinary School. By 1956, after reading a paper by Raymond P. Ahlquist, he had clearly formulated the aim of discovering a *selective* adrenaline receptor antagonist.

Raymond P. Ahlquist at the Medical College of Georgia speculated that there were two types of adrenoceptors (adrenergic receptors) as early as 1948 in order to explain the paradoxical actions of **adrenaline** (epinephrine) (figure 2.1) and **noradrenaline** (norepinephrine) on the cardiac muscles. He termed them alpha-adrenoceptor and beta-adrenoceptor, respectively. They were so named because they are both modulated by adrenaline. Because Ahlquist's theory was so revolutionary at the time, he found himself having difficulty getting his carefully reasoned and thoroughly researched paper published. He later commented: "The original paper was rejected by the *Journal of Pharmacology and Experimental Therapeutics*, was a loser in the Abel Award competition and finally was published in the *American Journal of Physiology* due to my personal friendship with a great physiologist, W. F. Hamilton."[3] Regrettably, Ahlquist's hypothesis was largely ignored for the first 10 years after its publication until Black began to look into its applications in drug discovery.

FIGURE 2.1 Adrenaline and Jokichi Takamine. © Japanese Ministry of Posts and Telecommunications.

Ahlquist's dual receptor theory stimulated Black to look for drugs with beta-receptor-blocking properties in a systematic way. In that sense, Black was the first to realize that the development of clinically useful selective inhibitors of adrenoceptors might introduce a new principle in drug discovery. In 1958, Black went to the Imperial Chemical Industries (ICI) Pharmaceutical Division in Alderley Park, Cheshire, and tried to lobby the company to work on beta-receptor-blocking drugs. Evidently convinced by his argument, ICI ended up hiring him as a senior pharmacologist to run the project even though he had no previous training in pharmacology. Together with chemist James Stephenson, Black led a team and began investigating how these messengers and receptors might be manipulated to produce a desired medical result. After a decade of research, they developed what came to be known as beta-blockers—substances that beat adrenaline to its target and, by acting as false messengers, occupy and block the receptors to bring about the intended pharmacological effects. In 1962, Stephenson and his colleagues succeeded in making a beta-blocker **pronethalol**. Unfortunately, pronethalol was withdrawn from further development when it was found to cause thymic tumors in mice. ICI eventually in 1964 discovered the drug **propranolol** (trade name Inderal), which possessed a better efficacy and safety profile. Propranolol is now widely used in the management of angina, hypertension, arrhythmia, and migraine headaches. Two additional beta-blockers, **atenolol** (trade name Tenormin) and **practolol** (trade name Dalzic), were later discovered and marketed by ICI. Beta-blockers were not just a new class of drugs; they represented a revolutionary approach to pharmaceutical research. Black changed the process of drug discovery from one of hunting to one of engineering by employing *rational drug design* to discover novel compounds that nature had not thought of.

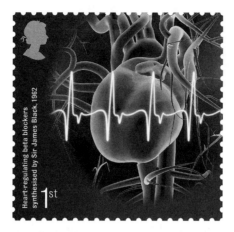

Heart-regulating beta blockers
synthesised by Sir James Black, 1962

1st

FIGURE 2.2 "Heart-Regulating Beta Blockers Synthesised by Sir James Black, 1962." © Royal Mail.

His monumental contribution garnered a Nobel Prize for him in 1988 and a commemorative postage stamp in 2010 by Royal Mail. However, as a chemist, I take issue with the design of the stamp (figure 2.2), which claims "Heart-Regulating Beta Blockers Synthesised by Sir James Black, 1962." For sure, without Black's genius idea to tackle beta-adrenoceptors selectively, there would be no beta-blockers. However, Black was a pharmacologist, not a chemist. It was his chemist colleagues who synthesized heart-regulating beta-blockers.

Histamine, Histamine Receptors, and Ulcers

Soon after the discovery of propranolol, Black shifted his interests in the early 1960s from beta-adrenoceptors to the effects of **histamine**. In an analogy to beta-receptors, Black hypothesized that the human body has two distinct types of histamine receptor cells: H_1 and H_2 receptors (today, we know that there are actually at least four types of histamine receptors: H_1, H_2, H_3, and H_4).

By selectively blocking H_2 histamine receptors, Black expected to obtain a drug to treat ulcers (H_1-receptor blockers are known as antihistamines; see chapter 4). In order to understand how a selective H_2 receptor works, we have to first take a look at the histamine molecule, as well as early antihistamines.

Histamine was first synthesized in 1907 by Nobel Laureate Adolf Windaus (figure 2.3) in Göttingen, Germany, from the decarboxylation of **L-histidine** (His or H), a naturally occurring amino acid. A few years later, the compound was found to be identical to a component isolated from animal tissue—thus it was named histamine, derived from the Greek *histos*, meaning tissue. Although histamine is found throughout body tissues and plays a key role in the regulation of various physiological processes, histamine itself is not a suitable drug because it modulates too many targets and has too many wide-ranging physiological effects. Sir Henry H. Dale (figure 2.4) of England

FIGURE 2.3 Adolf Windaus. © Posta Uganda.

FIGURE 2.4 Henry H. Dale. © Guyana Post.

extensively studied histamine's pharmacological actions.[4] He noted that it caused smooth muscles to contract and lowered the blood pressure. He also observed that in various species of animals, histamine's effects bore a striking resemblance to anaphylactic shock. Dale was awarded the Nobel Prize in Physiology or Medicine in 1930.

The first antihistamine (the substance that counteracts the effects of histamine) was discovered by French pharmacologist Daniel Bovet (figure 2.5) in 1937.[5] Between 1937 and 1941, Bovet conducted more than 3,000 experiments to find the chemical formulas upon which most of the antihistamines now prescribed are based. Antihistamines are effective in treating a wide range of allergic reactions. His discovery aided in the development of the first antihistamine drug, **phenbenzamine** (Antergan), for humans in 1942. In 1944, one of Bovet's own discoveries, **pyrilamine** (Neoantergan), was marketed as a drug. He did not patent it and did not make a penny out of his important discovery. But

FIGURE 2.5 Daniel Bovet. © St. Vincent and the Grenadines Post.

life was fair after all: Bovet won the Nobel Prize in Physiology or Medicine in 1957.

The two Nobel Laureates Dale and Bovet made tremendous contributions to the understanding of histamine's pharmacological actions. However, it was Leon Popielski who first showed that gastric acid secretion in the stomach could be stimulated by histamine. The gastric acid (hydrochloric acid, or HCl), a major constituent of the secretions of the gastrointestinal tract, is essential to the food digestion process. However, too much secretion of gastric acid was thought to cause heartburn and ulcers. Popielski made the landmark discovery as a student of Ivan Pavlov, who famously discovered "classical conditioning." In Pavlov's laboratory, Popielski made his own mark in unraveling some of the central nervous system's influence on gastric (stomach) function in dogs. Thanks to Popielski's work, we now appreciate that secretion of gastric acid by parietal cells of the stomach is initiated by the thought, sight, smell, or taste of food and is mediated by the autonomic nervous system.

By 1948, some of the actions of histamine were already well known. For instance, contraction of smooth muscles could be blocked by antihistamines—although those antihistamines were ineffective against histamine-induced gastric acid secretion or histamine-induced increase in the beating rate of the isolated guinea pig heart. The concept of histamine receptors was proposed in the 1940s by Heinz O. Schild, professor of pharmacology at University College London who, in 1966, also coined the nomenclature H for the receptors where the antihistamines acted. Bjorn Folkow and George Kahlson at Gothenburg University in Sweden investigated the vasodilating effect in the hind legs of cats and dogs. They showed that the increasing effect of higher doses was only partially inhibited by antihistamines used to treat allergic conditions. Based on those observations, they concluded, "The relative activities of several histamine analogues support the differentiation of histamine into at least two classes."[6] That was the first time the concept of *two receptor substances* for histamine appeared in the literature.

For histamine receptors, the H_1 type occurs in the nasal passages, other parts of the respiratory tract, and elsewhere in the body. Antihistamines generally target only the H_1 receptor. The H_2 type occurs in the billions of parietal cells located in the lower part of the stomach wall, where they produce hydrochloric acid. While the H_1 receptors were blocked by the antihistamines, there were no antagonists available for the H_2-receptor-mediated effects before Black's discovery. In the stomach, histamine stimulates gastric acid production by interacting with the H_2 receptors of the parietal cells. H_2-receptor antagonists, in turn, would be able to reduce the volume and acidity of gastric fluids by blocking the H_2 receptors.

Peptic ulcers—the word "peptic" is derived from a Greek verb meaning "to digest"—occur with about equal frequency among all socioeconomic groups, yet with varying incidence from country to country, and even within a country. In the United States, for example, about one person in 10 will suffer at some time from a small, usually sharply circumscribed sore that appears in the stomach (gastric ulcer) or the first section of the small bowel (duodenal ulcer). Duodenal ulcers are about four times as common as gastric ulcers. Most duodenal ulcers are associated with an excessive production of stomach acid, and gastric ulcers usually are not.[7] Many doctors at times considered gastric and duodenal ulcers as different disorders and presumed they had multiple causes. Duodenal ulcers were rare before 1900. Over the last 100 years, peptic ulcers have shifted from a disorder of the stomach in young women to a disease of the duodenum in middle-aged men.

The common belief that ulcers are caused by stress is now completely debunked by a solid association between ulcers and bacteria. We now know that a bacterium, *Helicobacter pylori*, is largely responsible for ulcers. But this fact was not known in the early 1960s; it was widely believed that the production of hydrochloric acid in the stomach was under the influence of histamine, but none of the antihistamine drugs at the time had a blocking effect on the secretion of gastric acid. Without the benefit of hindsight, Black decided to tackle the outcome (hydrochloric acid) rather than the cause (bacteria).

Black's Encore

Brilliant scientists are often not very good politicians. They have so much confidence in their scientific convictions that they are often too proud to conform to management's directions. James Black evidently belonged to that category. When Black presented his idea of pursuing selective antihistamines to the ICI managers, they said "no" because they wanted him to focus his energy on the development of the beta-blocker drugs, a more lucrative activity in the short term. They were correct in a way: Discovering a "me-too" drug is far cheaper with much higher chances of success. (A me-too drug is a drug that is structurally very similar to already known drugs, with only minor differences.) But Black had no enthusiasm for commercial development. In 1963, George Edward Paget, former ICI head of pathology, was hired to be the research director of the research group belonging to the industrial SK&F Laboratories Ltd., which had just moved to Welwyn Garden City. When Paget asked Black to recommend a pharmacologist to run biological research there, Black half-jokingly asked what was wrong with himself. So Paget hired him and wisely gave him a free hand to run his project.[8] Meanwhile, Paget also hired William A. M. Duncan away from his former employer, ICI, as the head of biochemistry to work along with Black. This proved critical for the birth of Tagamet, because Paget, and later Duncan who succeeded him, shielded Black from the interference and deadline pressure from SK&F's top management in Philadelphia and smoothed interactions between the aggressive Black and other team members.

In June 1964, following discussions with Black, Paget formally proposed that SK&F should invest in the discovery of a new type of histamine antagonist. He stated: "None of the known antihistamines are effective antagonists of histamine-stimulated gastric secretion...it is obviously worth testing close histamine analogues for antagonism to histamine-stimulated secretion."[9] Meanwhile, a team at the SK&F headquarters in Philadelphia, led by Virgil D. Wiebelhaus and under the guidance of Peter Ridley, was testing for anti-ulcer compounds by inducing ulcers in rat stomachs in vivo and looking to see whether a previously injected compound would protect them. The rivalry and competition between the two teams from Welwyn Garden City and Philadelphia would play out throughout the next decade.

Black once said, "The most fruitful basis for the discovery of a new drug is to start from an old drug."[10] But in this particular case for blocking peptic acid secretion, there was no old drug to start from. Histamine is the natural messenger, but it cannot be used as a drug. Ironically, peptic acid would chew it up. So his team began their chemistry using histamine as the prototype. At first, Black naïvely thought that it would be relatively easy for the chemists to design and synthesize H_2 antagonists. So he boldly told the management, "It won't take long and it won't need too many chemists involved."[11] He also predicted that they would have an inhibitor ready by Christmas. But it soon became evident that the project was much more complex than anticipated. So a student, Michael E. Parsons, was added to the team. By September 1964, the program was under way with Black and Parsons as pharmacologists, Duncan as biochemist, and Graham J. Durant as chemist. When Black drafted his former ICI team member Duncan onto his new team, he pointedly excluded several people formerly regarded as stars at the Welwyn Research Institute. There were bruised egos and hurt feelings among the veteran staff. Many left.

Christmas of 1964 came and went, and the team was not even close to getting an H_2 antagonist drug. Little did Black know that it would take many more years to achieve success.

Discovery of Tagamet

In drug discovery, no matter how great the pharmacology is, it always takes a chemist or a group of chemists to make the drug. It was no exception with the discovery of Tagamet. With Black's genius idea of tackling one of the two histamine receptors, H_2, it fell on the shoulders of medicinal chemists at the Welwyn Research Institute to make the compounds for drug candidates. Indeed, while Black was the father of the concept for the H_2-receptor blocker, it took a team of medicinal chemists to translate his idea into a drug.

The chemistry team began with a lone chemist, Graham Durant. But as the project's complexity became manifest, the chemistry team grew to three, Durant, C. Robin Ganellin, and John C. Emmett (figure 2.6), with Ganellin, a brilliant medicinal chemist, as the team leader. Ironically, a senior chemist from the Welwyn Research Institute was brought in to replace Ganellin as head of medicinal chemistry—a snub that almost caused his departure.[12] It was fortunate that Black and Duncan invited him back a bit later. Otherwise, the Tagamet breakthrough might have been crucially delayed or even quashed. The challenge was tremendous: The chemistry was unusual in that there was no known lead molecule (except histamine as the endogenous ligand) with the required properties to build upon. And neither natural products nor synthetic chemical compounds were known to be antagonists of H_2 receptors.

FIGURE 2.6 Drs. C. Robin Ganellin, John C. Emmett, and Graham J. Durant. © Robin Ganellin.

Ganellin was born on January 25, 1934, in East London and studied chemistry at Queen Mary College, receiving a Ph.D. in 1958 under Professor Michael Dewar. He spent a period in 1960 with Arthur C. Cope at MIT as a research associate. He then joined SK&F Laboratories Ltd. as a medicinal chemist in late 1960. Having been trained as an organic chemist, his knowledge of medicinal chemistry had to be acquired "on the job."

Between 1964 and 1968, Ganellin's team synthesized around 200 compounds chemically related to histamine and tested them for histamine antagonism. "Our aim was to find histamine analogs that bound to H_2 receptors but were sufficiently different from histamine not to stimulate gastric acid secretion," Ganellin later explained.[13] None of the 200 or so compounds showed any blocking activity in the assays. Four years of hard labor and no results—it looked as if the end of the program was near. Many doubts were expressed about the feasibility of blocking the action of histamine on gastric acid secretion. In 1966, John Emmett (who came to SK&F Labs Ltd. in 1965) synthesized **4-methyl-histamine**, which *stimulated* acid secretion. That was exactly the *opposite* of what they wanted to do, because 4-methyl-histamine was an H_2 agonist, not an antagonist. The silver lining was that they at least proved the existence of the second histamine receptor, establishing a clear target for drug research.[14] The discovery was scientifically monumental, but much, much more was needed to find a bona fide drug to suppress gastric acid secretion.

Understandably, there was enormous pressure within the company to abandon this approach in favor of the approach being carried out in Philadelphia. The scientists involved in the project were, however, firmly resolved to continue even after top U.S. management ordered the closure of the research program.[15] At the time, Duncan succeeded Paget as the director of research and development. He acted as the shield between Philadelphia and Welwyn.

He also changed the project name to "H_2 Receptor Program," so, at least on the surface, it seemed that Welwyn was not directly competing with Philadelphia's efforts to find a drug to treat peptic ulcers. In addition, Leon Greene, who later became SK&F's vice president of worldwide development, was so impressed by Welwyn's work that he decided to terminate the antispasmodic project in Philadelphia. Greene and Duncan worked together with great mutual respect, not always agreeing but never forgetting their ultimate objective of a new product.[11] Their leadership was vital in keeping the H_2 Receptor Program alive.

Later on in the program, the young and green pharmacologist Michael Parsons figured out a "minor" technical detail that would have a huge impact on the program.

Parsons was a student in pharmacology registered with Black to study for his Ph.D. One of his tasks was to assist Black with all the pharmacological assays. He noticed that Black had been making the same basic mistakes for the last four years. In his antagonism-binding assays, he had introduced histamine to the test medium only 30 to 40 percent of the maximum possible in real life. He did not do anything wrong per se; that was the textbook procedure that Black had chosen. But Parsons reasoned that 40 percent of histamine (the natural agonist) occupied too little of the receptors to have the antagonism to manifest. He then chose to flood the receptors with 100 percent of the possible histamine. Now, an H_2 antagonist would displace the portion of histamine already bound to the receptors. And as a consequence, the antagonism would be more pronounced.[15]

Parsons decided to use his modified assay procedure to test the 200 compounds that failed to show antagonism using the textbook procedure. One of the first six compounds that Durant synthesized before Christmas 1964 turned out to be positive. Walter Sneader, a renowned Scottish medical historian, called the episode "a cruel twist of fate!"[4]

Fascinatingly, Black was too impressed by the beta-blocker history in which the ring of **isoprenaline** was altered and the side chain was retained. Histamine also consists of an imidazole ring and a side chain. Black proposed that the imidazole ring should be modified and the side chain be retained. The chemists discovered that they had to keep the imidazole ring and modify the side chain to have good biological activity. Those first successful compounds were histamine analogues with an alkyl side chain terminating with a guanidine group. They subsequently found that analogues with longer chains were more effective at blocking gastric acid. The positive results electrified the team. Those were exciting days. The team saw each other every day. Although there were few formal meetings, every lunch time became a conversation about chemistry. Black was quite good at communicating with and directing chemists even though he did not understand the mechanics of chemistry. The many discussions they had together introduced the team to chemical questions of interest to pharmacologists and gave them a new insight into becoming medicinal chemists.

Durant's compounds had mixed activities, acting as both agonists and antagonists (also known as *partial agonists*). From 1968 to 1970, the team, grappling with difficult syntheses, investigated the structure-activity relationship (SAR) of the **guanyl-histamines**. They replaced the strongly basic guanidine group with non-basic groups. They focused on groups that are essentially neutral in water. Within a few days, John Emmett discovered isosteric replacement of a nitrogen atom on guanidine with a sulfur atom (an isothiourea fragment), which still had acid-releasing activity because it was basic and still acted as an agonist. Ganellin then studied additional non-basic groups, but these lost all activity. Eventually a thiourea was found that was weakly active as an antagonist and did not act as an agonist. Lengthening the side chain of this derivative and syntheses of substituted analogues led to the discovery in 1970 of **buri-mamide**, the first bona fide pure H_2 antagonist without agonist effects. Burimamide worked when injected into animals, and Black determined it to be pharmacologically selective enough for humans. Subsequently, there was an important meeting with the professor of gastroenterology from the University College Hospital London and his staff about how to proceed in testing burimamide in humans.

"No problem," they said after reviewing the safety data, "we can use medical student volunteers."[16]

"How soon?" Black asked.

"Well," they said, "first we have to get clearance from the Hospital's ethics committee."

"And how long will that take?"

"Well they have recently met, so the next meeting will take about six months."

"What, we can't wait six months; what can we do?"

"Well," they said, "if you were to volunteer yourselves, we do not need the ethics committee."

So Ganellin and Duncan immediately volunteered to be the first human guinea pigs to test burimamide. It was quite complex because they had to use a nasogastric tube so that acid could be collected from the stomach and titrated every 15 minutes. It was found that there was a reduction of gastric acid activity in their stomachs after they were given burimamide via injection. Duncan triumphantly reported to the Philadelphia headquarters that he had tested burimamide on himself. The only feedback that he received from Philadelphia was an official memo asking, "Did you check with corporate personnel about the insurance before you did it?"[16] His immediate reaction of choice words is left to the reader's imagination.

Unfortunately, burimamide was not bioavailable enough to be given orally. Being 10 times less active than the final marketed drug, cimetidine (Tagamet), it would have required a tablet of two grams! In order to improve the potency, the team tried another 70 new chemical derivatives before they discovered **metiamide**, which was 10 times more potent than burimamide and orally bioavailable. However, although it healed peptic ulcers in tests on humans, it was found to cause *agranulocytosis* (a dangerous depression of the production of infection-fighting white cells in the bone marrow). In essence, metiamide had efficacy but not safety. In June 1974, the decision was made to abandon metiamide. James Geddes, a future Tagamet manager, was in a marketing planning meeting in Amsterdam. He recalled: "The second day of the meetings, we got this call. They said metiamide had been suspended.... Well, God! I mean, we had to walk hand in hand over the bridge in Amsterdam because of the fear somebody might jump. Everybody was very depressed."[17]

The most likely cause of metiamide's toxicity was thought to be the thiourea group. Knowing that the thiourea group had historically been associated with a spectrum of toxicities, Bill Duncan had already wisely decided to seek a backup to metiamide even when it was looking very promising. By that time, new budgetary constraints were proposed. Convinced that they were on the right track, Duncan terminated as many of the other programs as possible to continue support for the H_2 project. The Welwyn chemists at the time turned their attention to non-thiourea analogues of metiamide. One of the thiourea replacements was cyanoguanidine, which has an amino-nitrile in place of the sulfur atom. The compound, synthesized by Graham Durant in 1972, was **cimetidine**. Cimetidine, with seemingly minor changes in the structure of the side chain in comparison to metiamide, was tenfold more potent than burimamide in vitro and had twice the activity of metiamide in vivo in inhibiting gastric acid secretion. It was evaluated for toxicology by 1973. It passed all trials with flying colors, and it did not cause the agranulocytosis that plagued metiamide. Cimetidine was shown to relieve symptoms and promote healing of lesions in a majority of patients with peptic ulcer disease at oral doses of 1 to 1.2 grams per day. It came in the form of pale green tablets or as a liquid, and it could be injected as well.

Cimetidine was marketed first in the United Kingdom in November 1976, and the FDA in the United States approved it in August 1977. The dosage varies according to the individual case, but in general, patients take one pill four times a day (q.i.d, or *quater in die* in Latin)—with each meal and at bedtime—for about six to eight weeks. However, peptic ulcers sometimes recur and therapy may be continued, depending on the circumstances in each case. By 1979, it was sold in over 100 countries under the trademark Tagamet, originating from "**antagonist cimetidine**." Tagamet revolutionized the medical management of peptic ulcer disease. It became the first blockbuster drug ever in medical history in 1986, generating over $1 billion in one year for SK&F.[18]

With the spectacular success of Tagamet, it became very important to discover an economic manufacturing process to make cimetidine. The production volume was as high as 1,000 tons a year, a large amount even by pharmaceutical standards. In retrospect, the need to have so much API (active principle ingredient) stemmed from the intrinsic low bioavailability of Tagamet, and it had to be taken four times a day. If a drug could be taken once daily (q.d., or *quaque die* in Latin; a drug taken twice a day is b.i.d., or *bis in die* in Latin; and a drug taken thrice a day is t.i.d., or *ter in die* in Latin), it would require much less API, a boon to the drug company in saving resources and time, and a boon to the patient who could take a pill only once a day.

The initial process used to prepare cimetidine, although adequate for initial supplies, involved a bottleneck step that required the reduction of an imidazole ester intermediate using lithium aluminum hydride (LAH). The LAH process was difficult, dangerous (it ignited when exposed to air), and expensive to operate, and SK&F was literally exhausting the world's supply of LAH for the production of cimetidine. To address these issues, process chemists Drs. Charles Berkoff and Elvin Anderson established a process research effort at the SK&F R&D facilities in Philadelphia aimed at finding cost-effective, practical, and patentable routes for synthesizing cimetidine. The most cost-effective method, using sodium in liquid ammonia for the reduction of the ester (the Birch reduction, a reaction taught in sophomore organic chemistry), was finally implemented and optimized. This cleaner, less expensive pathway helped cut tens of millions of dollars per year from the manufacturing cost. Many of the other innovative methods of preparing cimetidine also provided process patent protection and therefore enhanced its exclusivity in countries around the globe. This process patent strategy was especially important in Japan, where it was not possible to protect the product any other way.[19] This episode is a good example of how synthetic and process chemists can make tremendous contributions to the drug development process.

Never content to rest on his laurels, Black took the prestigious position of chair of pharmacology at University College London in 1972. This was the position that had been held by Heinz Schild (best known for the Schild plot used to determine how well a drug binds to the receptor [potency] and whether or not there is simple competition), who was retiring. Black was awarded the Nobel Prize in Physiology or Medicine in 1988, along with Gertrude (Trudy) Elion and George Hitchings, both from Burroughs-Wellcome Pharmaceuticals, for their discoveries of important principles for drug treatment. Later, he founded the James Black Foundation, a small biomedical research company with the aim of discovering prototype drugs. Sadly, Dr. Black passed away in 2010.

Drs. Ganellin, Emmett, and Durant were inducted into the U.S. National Inventors Hall of Fame in 1990 for discovering pharmacologically active guanidine compounds such as cimetidine (Tagamet). Their work was cited as having established a physiological role for histamine in the control of gastric acid secretions—the major cause of ulcers. Ganellin subsequently became vice president for research at the company's Welwyn facility. In 1986, following in Black's footsteps, he migrated from industry to academia. Ganellin was appointed to the SK&F Chair of Medicinal Chemistry at University College London, a position he still holds. He was also appointed a Fellow of the Royal Society in 1986.

In November 1997, the American Chemical Society and the Royal Society of Chemistry in the U.K. jointly recognized the cimetidine work as a milestone in drug discovery by designating it an International Historic Chemical Landmark during a ceremony at SmithKline Beecham's New Frontiers Science Park research facilities in Harlow, England.

§2.2. From a Frog into a Prince

Like the kiss from a beautiful princess that turns a frog into a prince, Tagamet turned SK&F into Wall Street's darling. Almost overnight, the company rose from a little-known drug firm in Philadelphia to being a major pharmaceutical company in the world. By 1986, SK&F ranked as the ninth biggest pharmaceutical company in the world in terms of sales. The development of Tagamet, appropriately dubbed as a "wonder drug," revolutionized the treatment of duodenal and gastric ulcers and other disorders (such as heartburn associated with acid reflux) that require a reduction in the amount of gastric acid secreted in the stomach.

Immeasurable Benefits

Before the introduction of Tagamet, treatment of peptic ulcers relied on extensive bed rest, the imposition of a bland diet, treatment with antacids and/or anticholinergics, and often surgery if the ulcers recurred.

But in 1978, the first complete year after the introduction of cimetidine into the U.S. market, there was a truly precipitous decline in ulcer surgery. The number of ulcer operations had dropped from 136,000 in 1966 to 97,000 in 1977, and then to 69,000 in 1978. It was the largest single decline of the decade. The number of operations rose to 81,000 in 1979.[20]

Cimetidine is not a cure, but it can produce rapid and sustained relief of ulcer pain, and it hastens the healing of ulcers after about a two-month course of therapy. Many doctors, judging from their own clinical experiences, say they believe that the drug has accelerated the decline in the need for ulcer surgery.

In 1980, Tagamet whisked past Valium, "mother's little helper," to become the world's largest-selling prescription drug at the time.

After the introduction of Tagamet in the United States, there was a 30 percent decline in the number of ulcer operations in this country over the preceding 12 years, according to a study reported in *The Lancet* by Harvey V. Fineberg and Laurie A. Pearlman of the Harvard School of Public Health in Boston. The World Health Organization (WHO) listed Tagamet as one of the world's most essential drugs for its ability to heal stomach ulcers without surgery.[21]

Windfall

With the innovative medicine came a windfall for Smith Kline & French. Tagamet became the first blockbuster drug in pharmaceutical history—it generated more than $1 billion per year in sales after 1986 and transformed SK&F into the ninth biggest pharmaceutical company in the world with regard to sales.[20] It was the top-selling prescription product in the United States, Canada, and many other countries. Years after Tagamet's introduction, patients in the United States spent more money on it than on any other drug, even the widely used tranquilizer **diazepam** (Valium). Some 15 million people in the world took Tagamet within two years; it was a runaway commercial success virtually everywhere it was sold. The drug accounted for more than one-third of SK&F's $1.77 billion sales in 1980 throughout the world.[21]

Introduced in the United States by SK&F in 1977, Tagamet was the pioneer acid blocker. Worldwide, it has earned the company a total of $14 billion and was the first drug ever to chalk up $1 billion in sales in a single year. But in the late 1980s, anticipating the worst when its Tagamet patent would expire in 1994, SK&F began conducting clinical trials and seeking FDA approval of an over-the-counter version of Tagamet. The wisdom of that decision became evident when Tagamet sales plummeted from $600 million in 1993 to only $400 million in the year after the drug lost its patent protection in May and became vulnerable to competition from less expensive generic varieties.

One of the most formidable competitors of Tagamet was Glaxo's Zantac.

§2.3. Zantac and More "Me-Toos"

Despite Tagamet's medical benefit to patients and its huge financial success for SK&F, it is not an ideal anti-ulcer agent. The half-life for Tagamet is only two hours, which means that it has to be taken four times a day. Normally, any medicine requiring more than a once-daily dosage often results in lower patient compliance. Moreover, it can cause minor skin rashes in some patients. An anti-ulcer drug with a longer half-life and fewer side effects would be more desirable.

Glaxo came up with ranitidine and sold it under the trade name Zantac in 1983. Zantac is more potent than Tagamet. It is longer-lasting with a half-life of three hours and needs to be taken only twice a day. It is also very safe, and not surprisingly, Zantac rapidly overtook Tagamet as the best-selling drug on the market.

Zantac, "Me-Too" but "Me-Better"

The story of the discovery of Zantac is just as fascinating as the story of Tagamet.

Zantac was discovered in the summer of 1976 at a small U.K. drug concern, Allen & Hanburys Ltd. (A&H). Initially a Quaker apothecary, A&H was established in 1715 in London, and it became a division of Glaxo P.L.C. in 1958. The company's fate was emblematic of the U.K. pharmaceutical industry.

In 1961, Allen & Hanburys hired David Jack as research director. Interestingly, Jack had been deputy research director at SK&F before he went to A&H in 1961, and he was the one who had hired Ganellin at SK&F. Under Jack's scientific leadership, A&H would go on to discover many important medicines, with Zantac as its pinnacle of achievement.[22] Jack set up multidisciplinary groups of chemists and biologists with defined therapeutic objectives. His favorite mantra was "ask the receptor questions." With Jack's superb managerial skills and riding on the wave of scientific breakthroughs, A&H/Glaxo was one of the most productive research organizations in the pharmaceutical industry. In the 1960s, they came up with three beta-blockers: **albuterol** (trade name Ventolin) for the treatment of asthma; **labetalol** (Trandate) as an antihypertensive; and **salmeterol** (Serevent), also for asthma. In the 1970s, Jack's group successfully discovered, developed, and marketed a serotonin inhibitor, **ondansetron** (Zofran), as an antiemetic, and a serotonin agonist, **sumatriptan** (Imitrex, a serotonin receptor [5-HT$_{1B1D}$] agonist), for the treatment of migraines. Although those important drugs made a lot of money for Glaxo, they were dwarfed in comparison by the next blockbuster drug, which blew all of them out of the water in terms of popularity and financial profit.

One of the most curious facts about SK&F's discovery of Tagamet was their remarkably nonsecretive publication of their preliminary results. Black insisted on publishing as early as possible—prematurely according to most companies' standards. He had his own rationales. On one hand, they staked out their territory and warned competitors. On the other hand, because the topic was so hot, scientists in both academia and industry jumped on the bandwagon, which provided invaluable resources in elucidating the mechanisms of H$_2$ receptors and issues around their early compounds such as metiamide and burimamide.

In 1972, David Jack and his pharmacologist colleague Roy Thomas Brittain attended a lecture by James Black at Hatfield Polytechnic Institute, where Black revealed that burimamide not only inhibited histamine-induced acid secretion in animals and humans, but also worked on acid secretion following food ingestion. Electrified by the astounding science and its outcome, Jack came back and suspended all other projects associated with mechanisms that the Peptic Ulcer Project Group was working on. Instead, he redirected the resources to focus on the H$_2$ antagonists with Roy Brittain (pharmacology) and Barry Price (chemistry) as the team leaders. True to Black's teaching, "the most fruitful basis for the discovery of a new drug is to start from an old drug,"[10] Glaxo's starting point was burimamide and later cimetidine (Tagamet). Their initial goal was to make a drug as good as Tagamet; as luck would have it, they made a drug that was even better than Tagamet.

The chemistry team consisted of John Clitherow, Lena Elliston-Ball, David Bays, and Roger Hayes, and John Bradshaw joined the project later.[23–25] At the time, it was already known that other heterocycles such as pyridine and thiazole could be used effectively in place of imidazole. Bays and Hayes tried to change the core structure of imidazole to tetrazole while incorporating the side chain of SK&F's burimamide. One of the reasons that Bays and Hayes chose tetrazole to replace imidazole was because they already had considerable experience with tetrazoles. Meanwhile, Clitherow, a senior medicinal chemist in the group, chose to use furan to replace imidazole because he had previously prepared furan derivatives in making analogues of the cholinergic agent muscarine during his Ph.D. research. Those choices confirmed an interesting point about discoveries; novel ideas need four elements to flourish: time, place, person, and culture. Indeed, scientists, chemists in this case, are considerably influenced by their experiences. Another case in point: When Bruce Roth at Parke-Davis began working on statins, he chose to use pyrrole as the core structure to replace Mevacor's decalin core structure simply because he had just worked on it during his postdoctoral research. The ultimate outcome was Lipitor, the best-selling drug ever.

Jack, Clitherow, and colleagues at A&H/Glaxo struggled for four years in futile attempts to circumvent SK&F's patent on imidazole with *non-basic* heterocycles. Neither Bays and Hayes's tetrazoles nor Clitherow's furan analogues were as potent as cimetidine. They were mostly weak inhibitors of the H_2 receptor. In other words, close, but no cigar. By 1976, the team was still not able to make a drug that was as good as cimetidine. The management became increasingly antsy and argued strongly to abandon the H_2 antagonists in favor of effective anticholinergics as a preferred mechanism for inhibiting acid secretion. Only three chemists were left to work on the project after most resources were transferred to work on anticholinergics.[25]

A breakthrough came when John Clitherow recalled an old reaction that he did during his Ph.D. research. In order to improve his furan analogues' water solubility, he thought of a reaction called the Mannich reaction, named after German chemist Carl U. F. Mannich (1877–1947). Simply treating furan with formaldehyde and dimethylamine created dimethylaminomethylfuran (also known as the Mannich base). Voila! Now the molecule still retained the furan framework but had a basic moiety outside the furan ring, a key attribute for the boosted aqueous solubility. The first analogue they made was incorporating the side chain of metiamide, but the molecule was only a moderately potent H_2 antagonist. When SK&F released their results on cimetidine (Tagamet), Glaxo immediately incorporated cimetidine's side chain to their dimethylaminomethylfuran scaffold and arrived at **AH18801**. Although the drug AH18801 was almost as potent as cimetidine, it had two problems. One was that it had poor crystallinity, and the other was its low melting point. Those dual drawbacks made the development chemists' job a nightmare, as opposed to a drug that was highly crystalline. The Glaxo chemists began to tackle the variation of the side chain in order to improve the physical properties.

Again, SK&F's advancements in the field became useful to Glaxo's endeavor. At one point, SK&F tried to replace cyanoguanidine with 1,1-diamino-2-nitroethene (also known as nitrovinyl), which had never been used in a drug. They incorporated the highly unusual nitrovinyl motif into their imidazole core structure. Unfortunately, the resulting drug showed no significant improvement over cimetidine in both animals and humans. So SK&F did not develop the compound further although it was patented. After reading the SK&F patent, Glaxo chemists attached the nitrovinyl fragment into their own dimethylaminomethyl-furan scaffold in an attempt to reduce hydrophilicity. To their pleasant surprise, although the resulting drug AH19065 was oil, it fortuitously tested as 10 times more potent in rats, and it was also less toxic than cimetidine. The drug, first prepared in early August 1976, would become **ranitidine** in time. But raniti-dine was a poorly soluble solid. Its hydrochloride salt prepared in 1977, which later became Zantac, was more soluble.[25] After extensive toxicity tests, clinical trials began in 1978. Because of the success of Tagamet, Glaxo had great confidence in the drug and threw a tremendous amount of resources behind its develop-ment. The clinical trials subsequently confirmed its efficacy and safety. Zantac was marketed in the United Kingdom in 1981 and in the United States in 1983, the fifth year when Tagamet had been in the U.S. market. To promote sales in the United States, Glaxo entered a joint promotion agreement with Hoffmann-La Roche, Inc., increasing Zantac's U.S. sales force to 1,200 from about 400.

Zantac was devoid of several shortcomings associated with Tagamet. The biggest drawback of Tagamet is that it has a short half-life and therefore has to be taken several times a day. But Zantac needs to be taken only once a day. In addition, Tagamet is not very selective, also binding to an important enzyme respon-sible for metabolizing drugs (cytochrome P450) and therefore having a propensity to cause drug-drug interactions with com-monly used drugs like **propranolol**, **warfarin**, **theophylline**, and diazepams.[26] Furthermore, Tagamet also binds to the androgen receptor. Its anti-androgenic action can cause sexual dysfunction and occasionally gynaecomastia in males. In reality, all of the aforementioned side effects were relatively rare and only took place after large doses, so they did not matter to the vast majority of patients who took Tagamet. However, Zantac, without those shortcomings, easily gained an upper hand in the marketplace.

In an audacious move, Glaxo set the price of Zantac as much as 50 percent higher than Tagamet and poured the extra revenue into a marketing campaign so effective that marketing executives still talk about it today. Glaxo promoted Zantac not just for ulcers but also for heartburn (or what the company preferred to call the more frightening "gastroesophageal reflux disease"). As a result of its good profile and Glaxo's all-out marketing efforts, after Zantac was launched in 1981, it overtook Tagamet in annual sales in 1987 to become the world's best-selling drug. In 1986, Zantac sales exceeded $2 billion, a blockbuster two times over.

In 1991, the Council of the Royal Society of the U.K. accorded David Jack and Roy Brittain, both biologists, the Mullard Award in recognition of their contribution to the discovery and development of drugs acting on adrenergic, histamine, and 5-hydroxytryptamine receptors. Having contributed to the discovery of four drugs in their careers, salbutamol, salmeterol, labetalol, and ranitidine, was a truly remarkable feat. However, in thinking about history, the author regrets that no chemists on the Glaxo team were awarded the same honor. After all, without chemists' designs and syntheses, there would be no drugs.

Ulcer Wars: Tagamet, Zantac, Pepcid, and Axid

With Black's revolutionary discovery of the histamine H_2 receptor and the success of Tagamet and Zantac, drug companies all over the world rushed into the fray, not unlike the California Gold Rush in the mid-1850s. Only three other companies managed to strike gold and take a drug to market: Yamanouchi, Lilly, and Teikoku. They were successful in discovering, developing, and marketing their own H_2-receptor antagonists as "me-too" drugs of Tagamet.

Yamanouchi, a Japanese pharmaceutical company, marketed the third histamine H_2 antagonist, **famotidine** (trade name

Pepcid in the United States). In the mid-1970s, it began to look for drugs to treat peptic acid.[27] Because the company had had much experience with prostaglandins, it initially focused on prostaglandins as possible anti-ulcer drugs. After having learned of SK&F's exploits with Tagamet, a team led by Isao Yanagisawa of Yamanouchi Central Research Laboratories began to search for their own histamine H_2 antagonists in August 1976. They replaced Tagamet's imidazole core with the guanidinothiazole ring. The guanidinothiazole ring used by Yamanouchi had been initially discovered by ICI Pharmaceuticals and was present in its H_2 antagonist **tiotidine**, which was not marketed due to toxic manifestations. Meanwhile, they also replaced Tagamet's cyanoguanidine side chain with carbamoyl-amidine. The resulting drug was the most potent histamine H_2 antagonist at the time. One problem with the drug was that it was not stable enough to be industrialized. After many failed replacements, a director of research in the company suggested that they try sulfamoyl-amidine. At first, team leader Yanagisawa was suspicious; not only were sulfamoyl-amidines notoriously unstable, they were very difficult to make as well. But the team went ahead and prepared the sulfamoyl-amidine, and it was stable. The compound would later become famotidine, which was 30 to 40 times more potent than Tagamet. Yamanouchi launched famotidine in Japan in 1985 with the trade name Gaster. One year later, Yamanouchi and its co-marketer Merck Sharp & Dohme marketed famotidine with the trade name Pepcid in the United States.

Eli Lilly & Company launched the fourth histamine H_2 antagonist in the United States, **nizatidine** (trade name Axid), an analogue of Zantac. Lilly was able to enter the market quickly because they had already had previous experience with antihistamines, and they enlisted the consultancy of a leading gastroenterologist, Mort Grossman. Reuben Jones, an eminent medicinal chemist at Lilly, laid the foundation of antihistamines in the 1950s during their quest for identifying an inhibitor of histamine-induced gastric acid secretion. Lilly's published work on histamine analogues was greatly appreciated and built upon by the SK&F medicinal chemists in their pioneering studies to discover burimamide. C. Robin Ganellin, the lead medicinal chemist for Tagamet, expressed his appreciation and called the outcome of Lilly's Axid "a poetic justice."[28]

In 1986, Glaxo's Zantac seized first place, with sales of $2 billion. From then on, Zantac outstripped Tagamet thanks to Zantac's superior efficacy and safety profile, as well as Glaxo's aggressive marketing tactics. And Glaxo charged a premium for Zantac, even though it was a johnny-come-lately in comparison to Tagamet.

The U.S. patent for SmithKline Beecham's Tagamet expired in 1994, Merck's Pepcid, 2000; Glaxo's Zantac, 2002; and Lilly's Axid, 2002, respectively. All four of them have generics on the market as cimetidine, ranitidine, famotidine, and nizatidine. They are now fantastically inexpensive and may be purchased over-the-counter. Ironically, SmithKline Beecham merged again in 2000 with Glaxo Wellcome, creating GlaxoSmithKline (GSK), the then-biggest drug company in the world. As a consequence, GSK now owns both Tagamet and Zantac, the top two H_2-receptor antagonists.

§2.4. Lessons

From the vantage point of posterity, many invaluable lessons may be learned from the saga involving Tagamet, Zantac, and other H_2-receptor antagonists in terms of science, business, and even the politics of drug discovery.

Undoubtedly, Tagamet, the first blockbuster drug in the world, owed its spectacular success to James Black's visionary scientific insight, innovation, and persistence. Proposing (although Black was not the first one to do so), researching, and confirming the existence of the H_2 receptor was itself a major scientific triumph. Working with medicinal chemists and translating the science into commercial success were even more impressive. Therefore, although science and the scientific approach are important for drug discovery, close and regular collaboration between chemists and biologists is essential. This may explain why drug companies prefer to station the chemists and biologists in geographical vicinity.

Even today, Black's wisdom still holds true: "The most fruitful basis for the discovery of a new drug is to start from an old drug."[10] The use of prototypes for drug discovery implies that compounds can be useful chemical tools for pharmacologists to help them unravel the mechanistic intricacies of particular physiological processes, often related to disease states, while acknowledging that few, if any, such compounds actually become medicines to be used therapeutically. History is replete with examples wherein incremental improvement of a prototype yielded a much better drug. Zantac was superior to Tagamet; Lipitor was better than Mevacor; ... and the list goes on.

With Black as the champion, SK&F adopted a daring approach of early publication and full discussion of problems, real and imaginary; this provided immeasurable benefits for science in general and for the discovery of Tagamet in particular. This was, of course, completely contrary to the conventional modus operandi of pharmaceutical companies, which always guarded their trade secrets under a shroud of secrecy, for good reason. But in this case, Black's strategy paid handsome dividends. Because the H_2 receptor was the state-of-the-art in pharmacology, scientists from both academia and industry eagerly jumped on the bandwagon, which provided almost unlimited resources in elucidating the unique mechanism. The knowledge generated by the whole scientific community supplemented the meager resources that SK&F had for the H_2 Receptor Program. In many ways, early publications accelerated the discovery of Tagamet. These days, drug companies guard their intellectual properties with the highest vigilance, which is certainly understandable. However, a fine balance needs to be struck to ensure that scientific advances is not smothered. Even academia has begun patenting their discoveries religiously, and those patents have generated millions of royalties for both the academic inventors and the universities. Isaac Newton said, "The reason why I can see far is because I stand on the shoulders of giants." In many cases, drug discovery builds on incremental advances in science, each scientist contributing a brick to build a scientific pantheon. Therefore, a reasonable publication strategy that discloses knowledge learned will not only help the scientific community, it will also help the industry itself. But this opinion may seem quaint and unpopular these days.

Persistence and tenacity are crucial for researchers to continue in the face of considerable difficulty and disappointment. It took over a dozen years for Tagamet to come to fruition after Black first initiated the project in 1964. The temptation to switch objectives when progress is slow should be resisted. With the benefit of hindsight, there were numerous junctures at which the fate of Tagamet might have turned out differently. Often, drug research and development takes time, with a little luck. Projects need appropriate amounts of time to mature, and constantly forcing projects too quickly through a company's pipeline may not be the best approach.

Regrettably, although SK&F had a huge lead on the industry in the field of H_2 antagonism, they failed to capitalize on their knowledge and human resources. They began the Tagamet development program in 1964, and 20 years later, there was not a second-generation compound past the advanced clinical stage

(Phase III). There was initially a backup compound called oxmetidine, but it was pulled out of development three months prior to marketing because of questionable reports of liver toxicity in South Africa. Several other compounds in development failed as well during late stages of safety assessments. The culture of the company did not seem to encourage innovation anyway, because soon after the launch of Tagamet, all key players involved in its development began to leave SK&F. After 10 years, all the key players were gone. At the time, SK&F was fittingly dubbed the "one-drug wonder."

To stave off the so-called post-blockbuster syndrome—a loss of growth and direction after the momentum generated by an extraordinarily successful product tapers off—SK&F merged with the Beecham Group in the United Kingdom. The "merger-as-equal" created SmithKline Beecham P.L.C. in July 1989. Thanks to a combined portfolio and combined sales force—the merged company had 52,000 employees worldwide even after laying off 5,000—SmithKline Beecham was able to keep Tagamet's blockbuster status until 1994 when its patent ran out in the United States. Beecham's drugs such as the antibiotic Augmentin and the antidepressant Paxil reduced the company's reliance on Tagamet. In 1980, Tagamet made up over a third of SK&F's total sales.[29] After Tagamet's patent expired, it was only 7 percent of the company's total sales by 1995. In order to fill the void that losing Tagamet created, SmithKline Beecham made a hostile acquisition of Sterling Winthrop from Kodak in 1994. Although a few symbolic managers from Sterling Winthrop were hired, the majority of the scientists from Sterling Winthrop were let go. That was the first major layoff done by a major pharmaceutical company in the United States. The notorious practice generated an outrage from academia. Many professors protested by refusing to send their best students to work for SmithKline Beecham, and the company suffered for almost a decade, failing to attract top talent. Sadly, these days, mergers and acquisitions have become commonplace in the pharmaceutical industry. Stripping assets with massive layoffs by the acquirers is so ubiquitous that few pharma companies are immune.

3

More Blockbuster Drugs for Ulcers

Prilosec, Nexium, and Other
Proton-Pump Inhibitors

Life is a struggle, not against sin, not against Money Power,
not against malicious animal magnetism, but against
hydrogen ions.

HENRY LOUIS MENCKEN *(1880–1956, American writer)*

§3.1. Better Ulcer Drugs Still Needed

In the first half of the 19th century, British physician William
Prout conclusively showed that gastric juice contains hydrochloric acid. U.S. Army officer William Beaumont examined the physiological control mechanism of gastric acid secretion by studying
Alexis St. Martin's chronic gastric fistula that had resulted from
a gunshot wound in the 1820s. Gastric acid is essential to digest
protein and emulsify fats. It breaks down food so it can go on to
the small intestine where nutrients are absorbed. Low levels of
gastric acid can contribute to a myriad of discomforts and diseases. On the other hand, too much of a good thing is bad. High
levels of gastric acid often result in heartburn and ulcers.

Heartburn is a symptom produced by reflux when digesting
food and gastric acid passes back up into the esophagus through the
sphincter muscle at the top of the stomach (see depiction of a stomach in figure 3.1). It is also known as GERD, for gastroesophageal

FIGURE 3.1 Stomach. © Brazilian Post.

reflux disease. If reflux occurs often and the body fails to sufficiently clear the acidic mixture back into the stomach, the tissue of the esophagus can be damaged, and that is when ulcers develop. Forty million Americans experience heartburn two days a week, and 60 million have it at least once a month. The disorder costs an estimated $10 billion in the United States, counting visits to doctors and hospitals, medications, and time lost from work, according to the American Gastroenterology Association.[1]

Before the emergence of Tagamet and Zantac as H_2 histamine-receptor blockers for the treatment of heartburn and ulcers, numerous medicines were available, but none were satisfactory. For over a century, heartburn sufferers had been taking over-the-counter (OTC) antacid products such as Alka-Seltzer, Maalox, Mylanta, Pepto-Bismol, Rolaids, and Tums. Most of them contain simple inorganic bases as the principal active ingredients. Americans alone spend approximately $1 billion a year on these antacids, which bring relief within minutes and work by neutralizing the stomach acid that causes heartburn. But because the stomach continues to produce acid, they remain effective for only a few hours at most. Sometimes, you have to take a quite lot of them to find relief from heartburn. For instance, it takes 60 grams of sodium bicarbonate ($NaHCO_3$) daily to neutralize acid for patients with gastric ulcerations whose stomach pH levels were 4.0. Meanwhile, calcium-based antacids like Tums and Rolaids can occasionally contribute to kidney stones, and aluminum- and magnesium-based ones like Mylanta and Maalox can sometimes be dangerous for people with kidney problems.[2] Therefore, H_2-receptor blockers, such as Tagamet and Zantac, are more powerful, longer lasting, and evidently superior to antacids because of smaller dosages (a couple of grams versus 60) and fewer side effects. Unfortunately, both antacids and H_2-receptor blockers may mask symptoms of bleeding ulcers among people with rheumatoid arthritis (RA). These patients often take the pills thinking that they will avoid the bleeding that can occur with high doses of RA drugs like ibuprofen. Although neither H_2 antagonists nor antacids cause bleeding, they may keep those with ulcers from recognizing the need to seek help. Tagamet may interfere with the body's ability to metabolize certain drugs, but the incidence, say researchers, is not significant (see chapter 2). Finally, H_2 blockers help only about 70 percent of sufferers; others need stronger drugs. Obviously, better drugs were still needed after H_2 blockers.

From the perspective of mechanisms of action, H_2 antagonists block only one of the three possible pathways for the activation of acid secretion: gastrin, acetylcholine, and histamine. They cannot completely suppress the hyper-secretion of acid. Furthermore, patients tend to develop a tolerance to Tagamet and Zantac, rendering them ineffective in a significant population of patients. Indeed, after taking them for a while, the efficacy seems to decline, possibly due to tolerance.

Then, in the early 1990s, Prilosec came along, a proton-pump inhibitor that is extremely effective at shutting down significant acid production in the stomach. Nearly 90 percent of patients who took the drug found complete relief, although it took a few days for it to take effect. A large majority of patients found that their symptoms could be eliminated by proton-pump inhibitors, which include Prilosec, Nexium, Prevacid, Protonix, and Aciphex.

§3.2. Prilosec

The discovery of Prilosec was another fascinating saga in medical science, with a combination of creativity, serendipity, and perseverance. Prilosec reduces gastric acid by blocking an enzyme called proton, potassium-ATPase (H^+,K^+-ATPase), also known as the proton pump. ATP, adenosine triphosphate, is a nucleotide that consists of an adenosine and a ribose linked to three sequential phosphoryl groups via a phosphoester bond and two phosphoanhydride bonds. ATP is the most abundant nucleotide in the cell and the primary cellular energy currency in all life forms. The primary biological importance of ATP rests in the large amount of free energy released during its hydrolysis. ATPases accomplish exactly that task: catalyzing the decomposition of ATP to give adenosine diphosphate (ADP) and a free phosphate ion. This dephosphorylation reaction releases energy. This process is widely used in all known forms of life.

In order to understand Prilosec, we must take a look at the molecular target: the proton pump.

George Sachs and the Proton Pump

In the 1970s, George Sachs at the University of Alabama in Birmingham showed that gastric H^+,K^+-ATPase functioned as the proton pump of the stomach.

George Sachs was born in 1935 in Vienna to physician parents.[3] The Sachses moved to Edinburgh in Scotland when George was three. In his youth, his father encouraged him to read extensively and bought five books each Friday for him to read during the next week. Until the age of 15, playing chess with his father was his only hobby and exercise. Incidentally, this author's only exercise in his youth was jumping (to conclusions).

George Sachs studied biochemistry and medicine at the University of Edinburgh, earning an M.B.Ch.B. degree (somewhat equivalent to a B.S. degree in the United States) in 1960. He then immigrated to the United States. After brief stints at Albert Einstein College and Columbia University in New York City, he started his independent academic career at the University of Alabama in Birmingham in 1963. There, Sachs disputed the conventional wisdom that gastric acid secretion was a *redox*-driven pump. Instead he proposed that the proton pump was rather an ATPase, which would be an optimal target for control of acid secretion. Although the revolutionary notion met with general disbelief, the chair of physiology at Alabama, Warren Rehm, was a staunch supporter.

As early as 1968, Sachs began a collaboration with a group led by Virgil D. Wiebelhaus at SK&F in Philadelphia to work on the treatment of peptic ulcers.[4] They developed a screen strategy to measure a partial reaction of the proton pump, and they identified some lead structures. However, the introduction of Tagamet by James Black's team in 1973 (see chapter 2) took all the wind out of their sails and eliminated all desire for SK&F to pursue another mechanism for the same indication. Also in 1973, Allen Ganser and John Forte at the University of California at Berkeley discovered the H^+,K^+-ATPase in the purified microsomes of a bullfrog's oxyntic cells.[5] Along with Ganser and Forte's group, the Sachs group established that potassium ATPase was the proton pump that moved acid across mucosa and gastric parietal cells. Sachs also hypothesized that the potassium ATPase proton pump might be a key drug target for the control of gastric secretion of acid.[6] However, Sachs's proton-pump target would not catch the drug industry's fancy until a chance encounter he had with a scientist from Astra Hässle in Sweden in 1977.

Astra Hässle's Gastrin Project

Hässle AB in Molndal, Sweden, was part of the Swedish drug firm Astra (now AstraZeneca). In 1967, Ivan Östholm, a research director at Astra Hässle, initiated a program in the gastrointestinal field.[7] It was known at the time that some local anesthetics reduced acid secretion when given orally to humans, and the local anesthetic lidocaine decreased acid secretion in experiments in dogs. It was speculated that lidocaine inhibited the acid-stimulating hormone gastrin locally in the stomach.[8]

Hässle called their project the "Gastrin Project" because they decided to tackle acid secretion by blocking gastrin. Professor Lars Olbe at the University of Göteborg, an expert in gastroenterology, was hired as an external consultant. Olbe would collaborate with Hässle's Gastrin Project from the beginning to the end when both Prilosec and Nexium were on the market.

During those days, drug discovery was still in a primitive stage—James Black's *rational drug design* model to modulate molecular targets, such as enzymes and receptors, did not become popular until after the mid-1970s. What the Hässle scientists did was the way drugs were traditionally discovered, namely, they made compounds and tested them on animals. This method required a large quantity of compounds, and the output was low. The advantage was that the end results were always clear—something either worked or it didn't. Furthermore, many drugs had been discovered that way. The Hässle Gastrin Project chose fistula Shay rats as their animal model.

They ran into problems immediately after the project got started.[9] The local anesthetics including **lidocaine** all contain basic amines. When they entered the stomach, they were protonated by the gastric acid and became inactive. Therefore, they needed to find an oral drug that was stable in the stomach. The chemists tweaked the structure of lidocaine by eliminating the basic amine and made many analogues whose structures were similar to that of lidocaine but without the basic amine, while still retaining the amide functionality. Since compounds with anesthetic properties also induced toxic effects in safety studies, they made carbamates in place of amide without anesthetic effects. Those carbamates were very effective at inhibiting gastric acid secretion in the rat model but did not work well in the dog model. As a rule of thumb in experimentation, the higher the species, the more similar and relevant to humans are the results. Therefore, the dog model carried more weight than the rat model. Slowly but surely, more compounds were made and more tests were done, and the Gastrin Project finally found a carbamate H81/75 (H stood for Hässle, 81 was the notebook number, and 75 was the page number) that was most effective in inhibiting acid secretion in dogs without detectable local anesthetic properties. Hässle tested the drug on human volunteers with high hopes, which were soon dashed: H81/75 had no desired effect at all for humans.

This was the peril of the traditional way of drug discovery. Not knowing what molecular target the drug was acting on made the outcomes decidedly unpredictable. In this case, the animal results did not translate to humans at all. Today, most, if not all, drug discovery programs target a molecular target such as an enzyme or a receptor. This contemporary approach is obviously not perfect either. While we may know how a drug interacts with the target, the results may not always translate into whole cells and the whole body. This is the reason why biomarkers have become such useful tools in predicting the efficacy and safety of drugs discovered using rational drug design.

In 1972, Hässle's Gastrin Project was restarted with a new approach using anesthetized dogs as an initial screening model, followed by conscious fistula dogs.[10] The team, led by veterinary pharmacologist Sven Erik Sjöstrand, consisted of zoo physiologist Gunhild Sundell; a brilliant medicinal chemist, Ulf Junggren; and five technicians. They surveyed the landscape of the anti-ulcer field; some significant progress had already been made by then, most noticeably by Searle in the United States and Servier in France.

Back in 1965, G. D. Searle and Company in Skokie, Illinois, was looking for an anti-ulcer drug by blocking the effects of gastrin, the acid-secreting hormone. Chemists at Searle used the structure of the C-terminus of gastrin as the basis ("prototype" in the parlance of medicinal chemistry) for synthesizing new compounds. They needed to synthesize the C-terminus tetrapeptide of gastrin, a peptide made of four amino acids. In December, Dr. Robert Mazur prepared **aspartylphenylalanine methyl ester**, a dipeptide, from two amino acids, and he gave it to his assistant James M. Schlatter to purify. To do the crystallization, Schlatter boiled the methanol solution with the dipeptide; when the flask was accidentally bumped, some of the powder got onto Schlatter's hand. Later, when he licked his finger, he noticed a very intense sweet taste. That was how he accidentally discovered the sweetener aspartame, now known as NutraSweet.[11] Although NutraSweet made tons of money for Searle, no anti-ulcer drug emerged from their efforts. The most advanced compound they discovered was antigastrin **SC-15396**, which was discontinued from development due to its toxicity.

Meanwhile, Servier in France prepared an antiviral drug, **CMN 131**, whose structure is very similar to that of SC-15396. CMN 131 was found to have acid antisecretory activities, but it too was discontinued from development because of its toxicity to the liver. Both Searle's SC-15396 and Servier's CMN 131 have one feature in common (i.e., the thioamide functional group). Hässle chemists chose CMN 131 as their starting point, hoping to remove the toxic group and increase the efficacy. In addition, they derived inspiration from James Black. During the discovery of Tagamet, Black and his colleagues at SK&F demonstrated that imidazole was important for acid control by H_2-histamine receptors.[6] The Hässle chemists wisely (though the wisdom was not apparent before the success of Prilosec, just like most other wisdom in life) selected a benzimidazole moiety to replace the toxic thioamide portion in Servier's CMN 131. This decision would prove critical if not prescient. The two blockbuster drugs that they later discovered, Prilosec and Nexium, both possessed the benzimidazole part, as do all other proton-pump inhibitors, including Prevacid, Protonix, and Aciphex. Benzimidazole also participates in the crucial catalytic cycle, which is described later.

It took a little over a year for the idea to come to fruition. Combining features in Servier's CMN 131 and SK&F's Tagamet gave rise to **H 124/26**, which was both efficacious and safe. At that time, the project was allowed to continue, but it could only expand in the chemical synthesis and biological test programs by collaboration with Abbott Laboratories in Chicago and with their financial support. Additional internal resources were apparently not available to the project. H 124/26 would have been a great compound to move forward, except that it was later found that a Hungarian patent already claimed the same compound for the treatment of tuberculosis. Hässle tried to negotiate with the Hungarian authorities but did not get anywhere because Hungary was still a communist country. The project looked hopeless and was almost shut down until they isolated **timoprazole**, which was not covered by the Hungarian patent. Timoprazole was H 124/26's oxidative metabolite isolated from dog urine because the H 124/26's sulfide was oxidized to the corresponding sulfoxide, which was even more active than H 124/26. Incidentally, Sankyo's pravastatin (Pravachol), launched in the United States in 1991 as an HMG-CoA inhibitor for lowering LDL cholesterol, was also an oxidative metabolite of mevastatin (compactin) isolated from dog urine. Mevastatin (compactin) did not reach the market because intestinal tumors were observed when it was given to dogs in large doses. With this encouraging data for timoprazole, the project was given a breath of new life and allowed to continue. Timoprazole was moved forward for toxicological studies in 1974. Unfortunately, it caused toxicities in the thymus and thyroid by blocking the uptake of iodine to the thyroid gland in rats. In addition to wrestling with toxicities for the thyroid and thymus, the Hässle group still had no clue about timoprazole's mechanism of action (MOA) and site of action, until a chance encounter with George Sachs in Uppsala.

Chance Encounter

In the summer of 1977, the Symposium for Gastric Ion Transportation was held in Uppsala, Sweden, in conjunction with the International Congress of Physiology. Hässle scientists presented their techniques and results using isolated guinea pig gastric mucosa. John Forte from U.C. Berkeley presented his data on the properties of the acid pump H^+,K^+-ATPase. Gaetano Saccomani from George Sachs's group at Alabama presented a talk titled "Transport Parameters of Gastric Vesicles."[12] During the presentation, they showed that H^+,K^+-ATPase was the proton pump of the stomach. They also described immunologic detection of the pump in different organs using a polyclonal antibody against ATPase, showing that it reacted strongly with the stomach and weakly with the thyroid.

After the presentation, Erik Fellenius, a biochemist from Hässle, asked Sachs whether this ATPase had been detected in thymus tissue. It had, but Sachs was sly, not wanting to give an answer without knowing the motive for the question. He responded that he would provide the information if Fellenius revealed the basis of his inquiry. After consulting with his colleagues and superiors at Hässle, Fellenius was given the green light. So the next morning, Fellenius told Sachs the story of timoprazole and its issues associated with the thyroid and thymus. In return, Sachs divulged the fact that the polyclonal antibody did indeed show reactivity with the thymus extract. The data were not presented in the symposium because he did not have a plausible hypothesis for the reaction of thymus with the antibody. As it so happened, Fellenius was in the process of going to the States to find a collaborator to investigate timoprazole's toxicities with the thyroid and thymus. Coincidentally, a Swedish postdoc, Thomas Berglindh, was working in Sachs's laboratories. Sachs was of course very eager to test timoprazole for acid secretion, especially considering his experience at SK&F. Fellenius visited Birmingham, and a collaboration deal was struck.

The Sachs group investigated if timoprazole was the ATPase inhibitor, but it was inactive. Just when they were scratching their heads, Fellenius told them that timoprazole was unstable in acid. Lo and behold, when tested under acid transport conditions, timoprazole was active.

In October 1978, Sachs and his postdoc Berglindh went to Hässle and reported their findings. They demonstrated that timoprazole and picoprazole (a close analogue of timoprazole) were prodrugs that were converted to active forms after accumulation in the acidic secretory canaliculus of the parietal cells. "Thus arose an entirely new domain of peptic ulcer therapy,"[3] acclaimed Sach's old friend Irvin M. Modlin, a professor at Yale Medical School.

Omeprazole

Having understood the MOA lent great impetus to the project because the structure-activity relationship (SAR) could be done in vitro in cells, rather than in vivo in animals. The pace was remarkably accelerated. Two senior medicinal chemists at Hässle, Ulf Junggren and Arne Brändström, led the chemistry team and carried out extensive SAR investigations. Gratifyingly, they discovered that they could separate the antisecretory activity from the toxicities associated with timoprazole by simply decorating both the pyridine and the benzimidazole rings with substitutions. Their pharmacologist colleagues quickly confirmed that the drug was indeed a proton-pump inhibitor as well. In 1976, they arrived at **picoprazole**, which had antisecretory activity in both rats and dogs but without toxic effect.

However, when given too much picoprazole for too long, three dogs developed necrotizing vasculitis in their small intestines.[13] Vasculitis is an inflammatory reaction in the blood vessels. This side effect prevented moving picoprazole to human trials, and thus the project was recommended for termination. A member of their scientific board pointed out that one of the three dogs in the control group not given the drug also had vasculitis, suggesting that the side effect might not be related to picoprazole (sometimes committees can be useful). It turned out that all three dogs were offsprings of one male dog, Fabian. Their vasculitis was a result of the antiparasitic drug given to them for intestinal worms. The project leaders later lamented: "A hereditary problem with Fabian delayed the project by several years."[8]

With the clouds over vasculitis toxicities cleared, picoprazole was then tested in humans and found to be the most effective antisecretory drug thus far, with a favorable safety profile.

Meanwhile, medicinal chemists kept optimizing the substitutions on the core structure. In January 1979, Ylva Örtengren was ready to make the final compound for the proton-pump inhibitor program. She should have used page 40 of laboratory notebook H 168 for that reaction. But her supervisor, Ulf Junggren, felt the compound that she was making would be "the one," and he wanted a code number easy to remember. So he suggested Örtengren use page 68, thus the compound had a code of H168/68.[14] Junggren's gut feeling was correct. H168/68 would eventually become **omeprazole**, with the later trade name Prilosec. Since **3,5-dimethyl-4-methoxy-pyridine** was known to increase the basicity of a pyridine ring, "in their enthusiasm to prepare a very potent compound," they went ahead and had Örtengren prepare it without systematically exploring the SAR. However, "by a stroke of good luck" in combination with a fantastic intuition, omeprazole was tested with a higher in vivo activity than any other combinations. As a matter of fact, omeprazole was the most powerful inhibitor of stimulated gastric acid secretion in experimental animals at the time. The drug had no sign of serious toxicity in animal models. By then, Hässle's Gastrin Project had gone on for over a decade. After omeprazole was synthesized, the original Gastin Project leaders Sven Erik Sjöstrand and Erik Fellenius left Hässle. Medicinal chemist Enar Carlson became the leader of the team, which included Björn Wallmark.

An investigational new drug (IND) submission was promptly filed, and human trials began in 1980. Astra Hässle's management challenged the team to develop the drug in record time. They had to overcome the issue of omeprazole's chemical instability, including both shelf stability and stability in the acidic environment of the stomach. The Phase I trials, led by Anders Walan, Christer Cederberg, and Lars Olbe, were completed uneventfully by 1983. As a matter of fact, the safety results generated great enthusiasm in the medical community around the world. Little did they know, new problems were just waiting around the corner.

In 1984, the Phase II clinical trials came to an abrupt halt. While the human trials were going on, Hässle also carried out lifelong high-dose treatment of rats to gauge omeprazole's chronic toxicity. Those rats developed enterochromaffin-like cell carcinoids (slow-growing types of tumors). But the team in Göteborg led by Enar Carlsson and Hellevi Mattson quickly determined that the tumor was caused by hypergastrinemia as a result of total inhibition of acid secretion and was not related to the safety of omeprazole.[14] Indeed, the same phenomenon could be reproduced using high-dose Tagamet as well. With that settled, omeprazole sailed through the Phase III trials. One of the studies involved physicians at 45 institutions in 13 countries who tested omeprazole and Zantac on 602 patients with stomach ulcers.[15] After one month of treatment, ulcers had healed in 80 percent of those getting omeprazole and 59 percent of those receiving Zantac, the brand name for the generic drug ranitidine. After two months, 96 percent of the omeprazole patients and 85 percent of the Zantac patients had healed. On the other hand, while proton-pump inhibitors such as omeprazole need a day or two to take effect, H_2 blockers such as Zantac work faster. Fast onset of relief is obviously an advantage of the H_2 blockers in comparison to the proton-pump inhibitors, an attribute still touted in today's TV commercials for the generic versions of the two classes of ulcer drugs.

In 1988, Swedish authorities approved omeprazole as the treatment for duodenal ulcers and reflux esophagitis. Astra Hässle sold it under the trade name of Losec. In the United States, Astra Hässle chose Merck Sharp and Dohme as their co-development partner. Merck had just had their first success with the licensed drug Pepcid, an H_2-histamine blocker also for ulcers, from the Japanese firm Yamanuchi. Pepcid quickly became a blockbuster, making $1 billion for Merck in 1985, and quickly squelched Merck's NIH (not-invented-here) syndrome. Flush with Pepcid's success, Merck CEO Roy Vagelos was eager to deal. Merck handled omeprazole's clinical development and new drug application (NDA) submission to the FDA. Omeprazole was approved in 1990, and Merck sold it with the trade name Prilosec.

How Does It Really Work?

Although omeprazole was determined to be a proton-pump inhibitor as soon as it was made in 1979, questions still lingered as to how exactly it worked on a molecular level. For instance, omeprazole behaves as a prodrug; it is not active in the absence of acid. Under the leadership of Per Lindberg, Hässle unlocked the mystery of the omeprazole cycle.

Lindberg knew he wanted to be a chemist when he was only 13 under his father's influence. In 1969, he became a graduate student at the University of Technology in Lund, Sweden. His Ph.D. supervisor, Börje Wickberg, assigned him the challenging task of isolating and identifying the toxic principle of the gray inky cap mushroom, *Coprinus atramentarius*. After one year of fruitless struggle, "in a more-or-less desperate experimental situation during present attempts to find the toxic principle, the author ate 300 g of boiled *C. atramentarius* and then, on the following day, took 20 mL of ethanol (40%, hint, whisky), but no uncomfortable effects were experienced."[16] It was fortunate that he did not do that experiment often because, later on, he found that the toxic principle coprine, which he eventually isolated and identified successfully, caused testicular lesions in both rats and dogs after a chronic dosage.

In 1974, Lindberg and Wickberg started a collaboration with Nobel Laureate Arvid Carlsson at the University of Göteborg to search for an alcoholic deterrent by capitalizing on their results with coprine. The project ended without a drug because of the chronic toxicities for animal testicles as mentioned before, but his achievement impressed one person enough to offer him a job. After hearing Lindberg's interesting talk at a symposium in 1976, Prof. J. Lars G. Nilsson at the University of Göteborg hired him as head of the Organic Synthesis Unit at the Department of Pharmacology. Nilsson's trust was well placed; Lindberg and his colleagues were highly productive in the central nervous system area. In addition, they also collaborated with Astra, filing approximately 10 patents in the field of dopamine and serotonin modulations.

Impressed by his ability, Hässle AB (a division within the Astra group) employed him as one of the section heads in medicinal chemistry in 1982. In addition to leading the efforts to look for omeprazole backups, Lindberg initiated an effort to elucidate omeprazole's MOA at the molecular level, in collaboration with biochemist Björn Wallmark and medicinal chemist Arne Brändström. They used many tools at their disposal including NMR (nuclear magnetic resonance), single crystal X-ray diffraction, and autoradiography of mice dosed with C-14-labeled omeprazole. The MOA was finally delineated, and it turned out that the drug worked through a cycle that they dubbed the "omeprazole cycle." The cycle is able to explain all the SARs. Many phenomena that once seemed idiosyncratic can now be completely explained. Here is how the omeprazole cycle works.

When omeprazole is accumulated in the acid space of the parietal cell, the nitrogen atom on the pyridine ring rapidly forms a spiro-intermediate, which is then converted to sulfenamide. The sulfenamide actually reacts with a thiol group of cysteine residues in the extracellular domain of the H^+,K^+-ATPase, and, therefore, the enzymatic activity is inhibited.

Dramatically, when Lindberg and Brändström gave a presentation of their exciting discovery in a medicinal chemistry symposium in Cambridge in 1985, Byk Gulden and SmithKline Beecham presented a poster with essentially the same mechanism.[16] The mechanism was so important that many companies were attempting to elucidate it. Partially thanks to the understanding of the MOA at the molecular level, the collaboration between Byk Gulden and SmithKline Beecham resulted in pantoprazole (trade name Protonix). Proton-pump inhibitors in general, and Prilosec in particular, revolutionized ulcer treatment. Prilosec brought a financial windfall to AstraZeneca, becoming a blockbuster shortly after its launch. At its peak in 1999, Prilosec's annual sales were over $6 billion, an equivalent of six blockbuster drugs.

§3.3. Nexium

Nexium was one of the most controversial drugs in the early 2000s. Many physicians and patients hailed Nexium as a godsend, as the sales figures show. But some vehemently denounced Nexium as a gimmick. Thomas A. Scully was one of them. Scully, administrator of the Federal Centers for Medicare and Medicaid Services, set overall payment policy for the programs. At a convention of the American Medical Association, Scully told doctors, "You should be embarrassed if you prescribe Nexium"[17] because it increases costs with no medical benefits. "The fact is, Nexium is Prilosec," Mr. Scully said. "It is the same drug. It is a mirror compound." But he also said, "Nexium is a game that is being played on the people who pay for drugs." "Mr. Scully was wrong in saying that Nexium and Prilosec were identical,"[17] said Dr. Joel E. Richter of the Cleveland Clinic, a former president of the American College of Gastroenterology. "Nexium is superior for some patients, particularly those with more severe forms of disease."

At first glance, Nexium *is* pretty much the same as Prilosec. As a matter of fact, Nexium is one of the enantiomers that Prilosec contains. Nexium is one-half of Prilosec. In order to understand the concept, we have to look at chirality.

What Is Chirality?

Chiral means *hand* in Greek. Chirality, denoting nonsuperposable molecules, is as old as life. But it was not realized until Louis Pasteur (figure 3.2) discovered it in 1848, 165 years ago. In preparing his Ph.D. thesis at the École Normale Supérieure, he noticed tiny crystals of the double salt ammonium sodium tartrate at the bottom of wine barrels. Interestingly, the crystals have two forms. Painstakingly, Pasteur separated the two forms of crystals by using a pair of tweezers under the microscope. When those two forms were dissolved in a solvent, they rotated the plane of polarized light in opposite directions: one to the left, the other to the right. When he mixed the two forms together in

FIGURE 3.2 Louis Pasteur and the tartaric acid crystals. © French Post.

a 1:1 ratio, they formed a racemic mixture, called racemate, and no longer rotated the plane of polarized light. Racemate is a 1:1 mixture of the so-called right-handed and left-handed molecules. The term "racemic" has its origin in the Latin for grape (*racemes*). Pasteur concluded that "the molecule of tartaric acid, whatever else it might be, is asymmetric in such a way that the image is not superposable."[18] In essence, Pasteur correctly deduced that the two forms behaved like left- and right-hand gloves. This was the first time anyone had demonstrated chiral molecules. He was merely 26 years old that year. How often does a 26-year-old create a new field of science?

Racemate contains two kinds of molecules that have the same constituent and the same shape but are mirror images of each other. In this respect, they are akin to a pair of hands, that have the same shape but, because of their mirror-image orientation, are distinct as right and left hands. These two types of molecules are known as enantiomers.

Upon hearing of Pasteur's great discovery, Jean Baptiste Binot, a celebrated French chemist and president of the École Normale Supérieure, was initially incredulous. He asked Pasteur to perform the experiment in front of his own eyes. The illustrious old man was visibly moved and seized Pasteur by the hand and said, "My dear son, I have loved science so deeply that this stirs my heart."[18] From then on, Binot became a staunch supporter of Pasteur's work and helped to catapult him into one of the greatest scientists of all time.

Again, the two forms that Pasteur separated are called enantiomers, which are spatial isomers that are also known as stereoisomers. They have the same chemical formula and structure but differ in their orientation in three-dimensional space. Enantiomers can be identified and distinguished by their optical characteristics. One enantiomer will rotate plane-polarized light to the right (called dextrorotary or [+]), and the other will do the same to the left (called levorotary or [–]). The dextrorotary molecule is sometimes called the right-handed molecule, and the levorotary molecule is sometimes called the left-handed molecule.

Is Nexium Really the Same as Prilosec?

Prilosec is a powerful drug for treating ulcers. But it does not work for everyone. This is because Prilosec has an idiosyncratic bioavailability. Hässle scientists were confident that they could easily fix the problem by improving the pharmacokinetics and metabolic properties of the drug. After all, they had accumulated a tremendous amount of knowledge about proton-pump inhibitors.

In 1987, a group led by Gunnel Sundén embarked on a focused program to find an omeprazole backup with better bioavailability. Their initial efforts at diversifying the core structure did not offer anything promising, so they decided to stick to the original core structure of omeprazole—little did they know how close the would-be backup would actually be to omeprazole.

Subtle modifications of omeprazole did indeed give two compounds that had some better attributes. However, only one had higher bioavailability than that of omeprazole. It was the S-(–)-enantiomer of omeprazole, which would eventually become Nexium.

Surprised? We should be. At first, Lindberg himself did not think of the isomers of omeprazole either. According to the omeprazole cycle, the two enantiomers would convert to an achiral intermediate under acidic conditions, and therefore the two enantiomers should behave the same. Two lucky breaks helped Hässle in arriving at the S-(–)-enantiomer of omeprazole.

One was the availability of enough material. When Lindberg and pharmacologist Lars Weidolf were looking into the pharmacokinetics of the two isomers, Sverker von Unge was able to isolate a few hundred milligrams of each isomer. When tested in rats, the *S*-isomer was four to five times more bioavailable than the other isomer (the *R*-enantiomer). Another lucky break was that they tested the *S*-isomer directly in humans and saw similar effects to what they observed with rats. Later on when the two isomers were tested on dogs, no significant difference of efficacy was detected for the two isomers. In this case, the rat was a better animal model than the dog. If they had used dogs for their initial in vivo tests, they probably would never have found Nexium.

Nexium was approved in Sweden in 2000 and in the States in early 2001, just when Prilosec's U.S. patent was about to expire.

§3.4. "Me-Too" Proton-Pump Inhibitors

Since the emergence of Prilosec, several "me-too" proton-pump inhibitors reached the market, and most of them became block-buster drugs. One of them is **pantoprazole** (Protonix), discovered by Byk Gulden (now a division of Altana Pharma AG) in Konstanz, Germany.[19] Byk Gulden had a rich history in developing antacids. Their magaldrate (Riopan), marketed since 1957, was a mixture of two inorganic bases: magnesium and aluminum hydroxides. They also developed a viable animal model, the Shay rat model, for gauging the efficacy of gastroprotective drugs. After the spectacular success of Tagamet and Zantac, Byk Gulden attempted to explore that MOA. Unfortunately, a search of the literature convinced their scientists that the H_2-histamine blocker field was pretty much optimized, and there was little patent space for them to squeeze into.

At the end of the 1970s, Byk Gulden pursued the mechanism of antimuscarinics, undoubtedly spurred by the success of Boehringer Ingelheim's **pirenzepine** (Gastrozepin). The active compound **zolenzepine** was obtained, but they were not able to eliminate the side effects of dry mouth and central nervous system–related blurred vision. During early 1979, Byk Gulden jumped into the field of proton-pump inhibitors. Using Hässle's timoprazole as their prototype, they strove to identify an analogue that was efficacious, safe, and outside the scope of existing patents.

That turned out to be hard to do. Initially, Byk Gulden's medicinal chemistry efforts were focused on modifying the benzimidazole portion of timoprazole. They synthesized some 160 compounds and identified **BY319** as their project compound. Unfortunately, BY319 caused enlargement and atrophy of the thyroid gland in rats and dogs, respectively. So it failed to reach clinical trials.

Not all was lost during the process; Byk Gulden isolated the key intermediate for the omeprazole cycle and elucidated the mechanism in 1985, the same time as the team at SK&F did.[16] This happenstance sowed the seeds of future collaboration between Byk Gulden and SK&F.

As the project proceeded, scientists at Byk Gulden realized that even minute changes in substitution would result in dramatic changes in stability and activity. They had to scrutinize their timoprazole-like analogues for all experimental data, such as activity, selectivity, stability in solution, solubility, and pharmacokinetics. The pyridine piece on pantoprazole was, unorthodoxly, not discovered by medicinal chemists. Rather, it was prepared by a small group of their scale-up chemists. When they were preparing BY319 in large scale, one of the byproducts was isolated from the rearrangement reaction of the corresponding pyridine N-oxide. With the "impurity" isolated, the chemists were able to prepare a series of analogues containing 4,5-dimethoxypyridine analogues, which contained pantoprazole.

In 1984, SK&F entered an agreement to co-develop proton-pump inhibitors. SK&F had a promising compound,

SK&F95601, which was about half a year behind the development of pantoprazole. In the end, pantoprazole was chosen because of its lower propensity to bind to CYP450, which made it less prone to drug-drug interactions. Moreover, because pantoprazole had such a good aqueous solubility and a very high stability in water, it became the first marketed proton-pump inhibitor for intravenous use in critical-care patients. It was approved by the FDA for marketing in the United States in 1994.[20]

Another "me-too" proton-pump inhibitor, **lansoprazole** (Prevacid), was discovered by Takeda Chemical Industries of Japan. In the United States, Takeda worked with Abbott Laboratories and established a joint venture called TAP Pharmaceuticals, the sole purpose of which was to sell Prevacid. Prevacid, marketed since 1995, had $3 billion in sales in 2001.

The third "me-too" proton-pump inhibitor, **rabeprazole** (Aciphex), was discovered by the Japanese drug company Eisai. It has been marketed in the United States since 1999.

Proton-pump inhibitors are widely used today. Some of them are sold OTC now thanks to their extraordinary safety profiles. Historically, Americans spent $13.5 billion on prescription proton-pump inhibitors in 2003, making them the second biggest-selling class of drugs after cholesterol-lowering medicines (statins such as Lipitor, Crestor, and Zocor). Prilosec alone generated more than $30 billion in sales since its introduction in 1988. Millions of patients suffering from GERD benefited from proton-pump inhibitors.

§3.5. The Saga Continues—Ulcers Are Caused by Bacteria!

In this day and age, it is not so easy to make groundbreaking discoveries. It was rumored that a U.S. Patent Office commissioner,

Charles H. Duell, once claimed in the early 1900s: "[E]verything that could possibly be invented had already been invented!"

The 20th century was marked by humans' triumph over infectious diseases. Whatever bacteria and viruses existed out there had been mostly identified at the dawn of the 21st century. Not surprisingly, when J. Robin Warren, a pathologist from Western Australia, "the wrong end of the wrong place," proclaimed that ulcers were caused by bacteria, few believed him.

In the 1970s, a novel technology called flexible fiberoptic gastroscope became available. The tiny microscope allowed close examination of the stomach's interior. In 1979, aided by a flexible fiberoptic gastroscope, Warren at the Royal Perth Hospital in Western Australia made a startling observation: He saw a large amount of curved and spiral bacteria in the vicinity of the stomach cells. Despite Warren's many attempts to interest his colleagues, most of them were indifferent—surely the bacteria are secondary invaders from outside of the stomach they thought. That was just common sense!

Only one of Warren's colleagues took an interest. Barry J. Marshall, a 29-year-old resident in internal medicine, was looking for a research project for his six-month rotation in the Gastroenterology Department where Warren was. Maybe because he was less familiar with the field's dogma, Marshall worked on Warren's project with gusto. He initially encountered many failures in growing the bacterium, just as Warren did, until serendipity came to the rescue.[21]

Normal bacteria usually grow in a couple of days, but Marshall was never successful within two days. In 1982, he took a long weekend off for Easter, but he forgot to clean up his Petri dishes—just like Alexander Fleming did with his discovery of penicillin. After Marshall came back to work, he examined the plates, and he found tiny, translucent pearls of bacterial colonies growing on the agar. One can only imagine the exhilaration he

FIGURE 3.3 Robin Warren and Barry Marshall. © Australia Post.

felt at the moment of discovery. Warren and Marshall (figure 3.3) later named the bacteria *Helicobacter pylori*. It is the organism that causes the most common chronic infection in the world. Medical science had ignored its existence for a century.

Robert Koch (1843–1910) (figure 3.4) was the German country doctor who discovered *Mycobacterium tuberculosis*, the

FIGURE 3.4 Robert Koch. © Deutsche Post.

bacterium that causes tuberculosis. For a bacterium to cause a certain infection, Koch had a triad of postulates: association, isolation, and retransmission. By 1982, Warren and Marshall had firmly established *Helicobacter pylori*'s association with ulcers, and isolation was also accomplished. What they needed was to establish that *Helicobacter pylori* was the cause of ulcers—retransmission.

That posed an ethical dilemma: One could not infect a healthy individual with a bacterium and see if it caused a disease (although the Nazis had no qualms about doing just that during World War II). In 1984, Marshall volunteered himself as the human guinea pig. He drank 50 mL of a broth laced with the *Helicobacter pylori* bacteria. Not only did he immediately become violently ill, he was later diagnosed with highly active gastritis. He rapidly developed ulcers that were swarming with the bacteria. After that experiment was published, all the naysayers became silent. The causation of gastric inflammation by *Helicobacter pylori* was thereby indisputably established.

In 2005, Warren and Marshall were awarded the Nobel Prize in Physiology or Medicine for their discovery of *Helicobacter pylori* and its role in gastritis and peptic ulcer disease.

Since Warren and Marshall's discovery, a reasonable treatment for ulcers often involves a combination of an antimicrobial to eradicate the bacterium, in addition to an H_2-histamine blocker or a proton-pump inhibitor.

4

Antihistamines as Allergy Drugs

...know now that drug discovery requires the same
characteristics of a medicinal chemist as of a partner in
marriage: patience, endurance, stubbornness and fantasy.
After all, there must be a reason why chemists in the United
States have the lowest divorce rate of any professional group.

GERHARD SATZINGER, *Parke-Davis Pharmaceuticals*

Most blockbusters have at least one thing in common—they are all widely prescribed to treat common illnesses such as hypertension, high cholesterol, pain, ulcers, and depression. Allergies (figure 4.1) are another malady that afflicts more and more Americans. To many, allergies are no longer an inconvenience but a major annoyance with constant sneezing and itching. For them, an allergy medicine is often needed to relieve the symptoms. As a consequence, many antihistamine allergy drugs, especially nonsedating antihistamines, have become blockbuster drugs.

§4.1. What Causes Allergies?

Allergies are the sixth leading cause of chronic illness in the United States. More than 50 million Americans have allergies and spend in excess of $18 billion a year on medical treatment.[1]

The word *allergy* was coined by Austrian pediatrician Clemens von Pirquet in 1906. According to his definition, allergy was manifest in cases of serum sickness, hay fever, sensitivities to

FIGURE 4.1 Asthma-Allergy. © Post Denmark.

mosquito bites and beestings, and various idiosyncratic food reactions, as well as in individuals who had been exposed to, or successfully immunized against, common infectious diseases such as diphtheria and tuberculosis.[2] Today, the word *allergy* is broadly associated with allergic rhinitis, asthma, hay fever, and food allergies.

Allergies are the malady of civilization. In ancient times, allergies like hay fever and food allergies were virtually nonexistent. The first report of a case of allergy did not appear until the 1870s in Europe. The beginning of the 20th century saw a sharp rise in allergies. Nowadays, hay fever is so prevalent in the United Kingdom, that there are 1.4 to 1.8 million students who are drowsy from taking antihistamines. U.K. educational authorities even schedule the exams away from the peak of pollen season. In the United States, hay fever is the number-one chronic disease. Until we learn how to turn off the genes responsible for hay fever and asthma, these afflictions will remain among the most irritating of our existence.

During evolution, humans developed the immune system to fight the real danger of foreign invaders like bacteria and viruses. It turns out that the human body has two types of responses toward tissue damage or infection. One is the innate immune response; the other is the adaptive immune response (as an example, "AIDS" stands for *acquired* immunodeficiency syndrome). As civilization progresses and human hygiene improves, there is less and less exposure to infective agents, especially in developed countries. When in contact with allergens like pollens, the human body mistakenly assumes that it is being attacked by infective agents and its immune system goes into overdrive and generates antibodies to attack normal cells. This often results in a runny nose, sneezing, and many other uncomfortable allergic symptoms. Ironically, poor people in developing countries are relatively devoid of allergies because they are routinely exposed to real germs on a daily basis.

How allergens work on the molecular level has been the subject of numerous investigations over the last decades. Our current understanding (figure 4.2) is that an allergen such as pollen invokes the body's immune system and T cells in the blood, tissues,

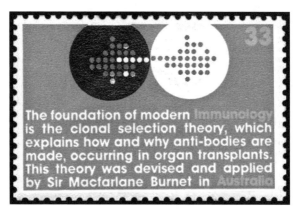

The foundation of modern Immunology is the clonal selection theory, which explains how and why anti-bodies are made, occurring in organ transplants. This theory was devised and applied by Sir Macfarlane Burnet in Australia

FIGURE 4.2 Immunology. © Australia Post.

and even in the lining of the nasal passages, producing substances that control the production of immunoglobulin E (IgE), a class of antibodies first identified in 1967.[3] Normally, a person produces only a small amount of IgE. When allergens come in contact with the nose, eyes, or lungs, they send false signals, alarming immune systems in people with allergies. Mistaking those allergens as invading foreign substances, the cells produce IgE too abundantly. When they are attached to the receptors on the surface of the mast cells, a copious amount of **histamine**, **leukotrienes**, and 15 different inflammatory molecules are released uncontrollably. These powerful molecules damage capillary blood vessels, wreak havoc on body tissues, and cause swelling, itching, redness, skin eruptions, sneezing, runny noses, and watery eyes.

Antihistamine drugs relieve allergic symptoms because the active ingredients are molecules that fit like keys into the locks, which are the histamine receptors H_1 and H_2 (see chapter 2 for more on the H_2 receptor) on the surface of mast cells in the capillaries, blocking the signals that cause allergies.

§4.2. Conventional Antihistamines

Before the emergence of antihistamines as effective treatments for allergies, many medicines were used with varying degrees of success. Opiates and asthma cigarettes containing stramonium or belladonna were the earliest remedies for allergy relief. From a chemical perspective, the principal ingredient in belladonna is alkaloid **atropine**, which is a nonselective muscarinic acetylcholinergic antagonist.

Adrenaline (epinephrine) and **noradrenaline** (norepinephrine) are secreted by the adrenal gland. These hormones bind to their corresponding receptors and elicit corresponding biological responses. Stimulation of the sympathetic nervous system due to *fright* leads to preparation of the system for *fight* or *flight*. Japanese chemist Jokichi Takamine at Parke-Davis isolated adrenaline in 1902. Adrenaline was often used in inhaled form to treat asthma in the 1920s in the United States. An increase in adrenaline results in the dilation of bronchi, dilation of pupils, and constriction of peripheral (outside the central nervous system) blood vessels, thus relieving asthma symptoms.

The 1920s also saw a widespread use of ephedrine, an active component in the traditional Chinese medicine *ma huang*. In the 1930s, even amphetamines were frequently used as a treatment for hay fever during hay-fever season. Also in 1930s, **theophylline** and **aminophylline** became especially popular in the United States as allergy medicines.[4] In fact, aminophylline had largely surpassed adrenaline as the drug of choice for treating acute asthma attacks at the time. It is interesting to note that both theophylline and aminophylline are **caffeine** that is demethylated at different positions. **Sodium cromoglycate**, a mast cell stabilizer that prevents histamine release, was also used in treating allergies during that period. These days, it is used only in a few countries for eye-related irritations.

None of these aforementioned early allergy drugs are as effective as antihistamines for the treatment of allergies.

Bovet's First Antihistamines

As mentioned in chapter 2, the molecule histamine, β-imidazolylethylamine, was first synthesized by German chemist Adolf Windaus in Göttingen in 1907 as a mere venture in chemistry. Windaus heated **histidine**, an amino acid, and obtained the product histamine after elimination of a carbon dioxide molecule. It was not until four years later in 1911 when British pharmacologist Henry H. Dale isolated histamine from the human body. Dale carried out extensive investigations of the

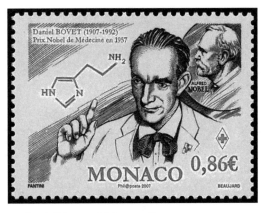

FIGURE 4.3 Daniel Bovet. © Monaco Post.

pharmacology of histamine, in addition to ergot alkaloids and tyramine. He won the Nobel Prize in Medicine or Physiology in 1936 for discoveries relating to chemical transmission of nerve impulses.

The first antihistamine was discovered by Swiss pharmacologist Daniel Bovet (figure 4.3) in 1937.[5] Bovet, born on March 23, 1907, in Neuchâtel, Switzerland, earned his doctorate in natural sciences at the University of Geneva. Between 1929 and 1935, he worked as an assistant in the Laboratory of Therapeutic Chemistry at the Institut Pasteur in Paris. Dr. Bovet became the head of the laboratory in 1936. By then, he had already begun his search for compounds that would relieve the sometimes fatal effects of asthma. He was fascinated by the fact that in nature, and in the human body, no product existed to counteract the excessive effects of histamine. Besides asthma, those effects included hay fever and hives. Bovet had access to a store of compounds that had been synthesized by the French physiological chemist Ernest Fourneau (1872–1949). These were the compounds that he tested to find the first leads to antihistamines. Between 1937 and 1941, Bovet conducted more than 3,000 experiments to find the compounds upon which most of the antihistamines now prescribed are based. His discovery led to development of the first antihistamine drug, **phenbenzamine** (Antergan), for humans in 1942, and in 1944, one of Bovet's own discoveries, **pyrilamine**, was marketed as a drug. Pyrilamine (trade name Neoantergan) is the U.S. name; it was synthesized at Rhône-Poulenc in Europe. The drug is still sold in its generic form as mepyramine. He did not patent it and never made a penny from his important discovery. Things are very different today, but in the past, scientists considered money earned from their discoveries as mere "icing on the cake." Bovet won the Nobel Prize in Physiology or Medicine in 1957.

How do histamines work? We learned in chapter 2 that there are at least four histamine receptors. The H_1 type occurs in smooth muscle, endothelial cells, nasal passages, other parts of the respiratory tract, and nearly elsewhere in the body. Antihistamines generally hit only the H_1 receptor. By inhibiting the histamine receptor, antihistamines block the pathways of allergic reactions. Antihistamines are antagonists of histamine receptors that displace histamine competitively from its receptors and block the effects of histamine. Histamine receptors, in turn, are G-protein-coupled receptors (GPCRs) that contain the typical seven-transmembrane loop motif. Other common GPCRs include calcium channel receptors, andrenergic α_1, dopamine D_2, serotonin $5\text{-}HT_2$, and muscarinic receptors. They have traditionally been productive targets for drug discovery. Many blockbuster drugs' mechanisms of action (MOAs) involve modulating GPCRs. Histamine receptor inhibitors alone include several blockbuster drugs such as famotidine (for ulcers), ranitidine (for ulcers), loratadine (for allergies), fexofenadine (for allergies), and cetirizine (for allegies). Moreover, many drugs such as fenoldopam, captopril, and prazosin that serve as dopamine, serotonin, angiotensin, and adrenergic receptor antagonists are marketed.

Indeed, nowadays, GPCR inhibitors account for 20 percent of the top 50 best-selling drugs and greater than 50 percent of all drugs marketed today.

The applications of the first-generation antihistamines discovered in the 1940s were limited since they caused significant adverse effects like sedation, memory impairment, and psychomotor dysfunction. Take Bovet's Neoantergan, the first antihistamine, as an example; it had a side effect of severe drowsiness. It is easy to understand why because its MOA is blocking the H_1 receptor, which also exists in the brain, and Neoantergan can cross the blood-brain barrier (BBB) with ease.

The second-generation antihistamines had significantly fewer central nervous system (CNS) adverse effects because they penetrated the BBB much less extensively.

Rieveschl's Benadryl

The most popular antihistamine is probably **diphenhydramine** (Benadryl), which is still a household name today. Beta-dimethylaminoethylbenzhydryl ether hydrochloride (Benadryl's full chemical name) has much less of a drowsiness side effect than Bovet's Neoantergan had. It was invented in the early 1940s by George Rieveschl, a chemical engineering professor at the University of Cincinnati.[6]

Born in 1916 in Ohio, Rieveschl studied commercial art at the Ohio Mechanics Institute.[4] After graduation in 1933, he sent out more than 200 job applications for work in commercial art. He received six responses, all rejections. He then turned his attention to chemistry, enrolled at the University of Cincinnati, and paid tuition of $35 a semester. There, he earned a bachelor's degree in 1937, a master's in 1939, and a doctorate in 1940. He stayed on at the University of Cincinnati as a chemical engineering professor.

Initially, Rieveschl, a very good organic chemist, aimed to prepare better muscle-relaxing medications, known as antispasmodic drugs. Working with his graduate student Frederick Huber, they prepared a series of compounds. One of them, A524, was found to be a powerful blocker of histamine instead. In 1943, Rieveschl went to work at Parke-Davis, then the nation's largest drug manufacturer, to test his discovery. Parke-Davis later bought the rights to A524, which in time became diphenhydramine and then Benadryl. Scientists were much more cavalier those days than we are today. Even before any clinical trials, a chemist at Parke-Davis took two pills of Benadryl before lunch, and he did not experience any discomfort after two servings of French fries. (He had had a serious allergy to potatoes for a long time before that.) His experience was repeated in the majority of patients who took Benadryl, and Parke-Davis started marketing it as a prescription drug in May 1946. Rieveschl received a 5 percent royalty for the 17 years before the patent expired. That made him a very rich man.

Benadryl has also been extended to other indications. The active ingredient of antiemetic and anti–motion sickness medicine Dramamine contains diphenhydramine, Benadryl's active principal ingredient (API). In 1947, G. D. Searle, a drug company in Chicago, sent Dr. Leslie Gray at Johns-Hopkins an experimental antihistamine drug called Dramamine, which was a mixture of Benadryl and chlorotheophylline (a derivative of the bronchodilator theophylline). Gray at first gave it to a pregnant woman, who was not impressed by its effects on allergies. But she noticed that her car sickness went away. Building on this serendipitous observation, in 1948, Gray gave Dramamine to seasick members of the American Olympic Team, who arrived in London refreshed without being seasick. Later on, Gray worked with the U.S. military and saw that Dramamine alleviated the seasickness of 94 percent of the troops. One year later, Dramamine began to be widely used by civilians as well. These days, Dramamine is still sold over-the-counter (OTC) at every drugstore.

The discovery of Benadryl was also significant because it was one of the first findings that specific histamine receptors in the capillaries can be affected by different compounds other than histamine. So there are now many antihistamines that can counter different histamine receptors. This development further promoted the receptor theory advanced by Paul Ehrlich at the beginning of the 20th century.

Benadryl, as the most popular antihistamine, is still in use today. Nearly everyone knows about Benadryl, but few people know that Benadryl was the prototype for Prozac. Indeed, as James Black said, "The most fruitful basis for the discovery of a new drug is to start from an old drug."[7] When Eli Lilly was searching for an efficacious and yet more selective and thus safer (much less toxic than the tricyclics) antidepressant, chemist Bryan Molloy and his colleagues used Benadryl as their starting point. He began to tweak the molecule by replacing the original substituents on Benadryl with different functional groups. Through a process called structure-activity relationship (SAR) studies, he found out that the N-methyl ethylamine moiety was necessary to maintain the pharmacological activity (pharmacophore), so he kept that side chain constant. In time, he manipulated Benadryl's diphenhydramine structure into the phenoxypropylamines, which unfortunately also inhibited another target, an acetylcholine receptor. In the end, one of the compounds selectively blocked the removal of serotonin while sparing most other biogenic amines. The more selective compound, which later became **fluoxetine** (Prozac), is a selective serotonin reuptake inhibitor (SSRI). Approved by the FDA for market in 1988, Prozac rapidly revolutionized the treatment of depression thanks largely to its safety profile. Prozac profoundly transformed debilitating depression into a manageable disease for many patients.[8]

Schering-Plough's Optimine

Schering-Plough Corporation in Kenilworth, New Jersey, had a long tradition of discovering, manufacturing, and marketing antihistamine drugs before it was acquired by Merck in 2009. The company was founded in 1921 as the American subsidiary of Germany's Schering A.G.

During World War II, Schering A.G. was treated as enemy property and suffered in terms of finance and resources. But it witnessed a spectacular postwar growth, especially in the fields of hormones and antihistamines. Schering's first antihistamine drug was **pheniramine** (Trimeton). The discovery in itself was not earth-shattering because it was merely a "me-too" drug of Benadryl, and its structure is virtually identical to Benadryl's with two minor variations: Trimeton replaced the oxygen atom with a carbon atom and switched one of the two benzene rings on Benadryl to a pyridine ring. Those two small changes seemed to do the trick: Introduction of the pyridine ring apparently made the drug less likely to penetrate the brain, and therefore Trimeton had fewer drowsy side effects in comparison to other antihistamines on the market at the time. Trimeton was widely used in cold medicine and made a lot of money for Schering.[9]

Interestingly, the presence of the pyridine ring would persist on virtually every single antihistamine that Schering brought to market, including chlorpheniramine (Chlor-Trimeton), azatadine (Optimine), loratadine (Claritin), and desloratadine (Clarinex). One could say that pyridine was Schering's favorite heterocycle, the same as thiophene was Eli Lilly's favorite heterocycle, which appeared in many of Lilly's CNS drugs.

Despite Trimeton's financial success, its half-life was short (one to two hours), and it thus had to be taken several times a day. Taking advantage of chlorine's resistance to hepatic metabolism, Schering's medicinal chemist Frank J. Villani installed a chlorine atom on the benzene ring of Trimeton. The result was **chlorpheniramine** (Chlor-Trimeton), which had all the enduring attributes of Trimeton but needed to be taken only once a day. Needless to say, Chlor-Trimeton enjoyed even better success than Trimeton before going went off patent in 1976.

Schering became Schering-Plough when Schering A.G. merged with Plough, Inc., in Tennessee in 1971. It could not afford to rest on its laurels of success in the antihistamine field for not only did Chlor-Trimeton saturate the market, its patent was to expire soon. Schering-Plough's scientists led by Frank Villani answered the challenge by coming up with **azatadine** (Optimine) in 1973. The structure of Optimine is different from that of Chlor-Trimeton, which is a floppy molecule. Optimine is a rigid version of Chlor-Trimeton. While Chlor-Trimeton has two rings, a benzene ring and a pyridine ring, Optimine has four rings, which restricted the flexibility of the molecule. Making a molecule rigid has many advantages. First of all, the molecule could be more potent by binding the receptor more tightly. Another benefit is that the new molecule is novel and can be patented with a new patent life. When Optimine was initially tested in cats, it showed little sedating effect; unfortunately, the results did not translate to humans, and it made humans tired in clinical trials.

Schering-Plough's Trimeton, Chlor-Trimeton, and Optimine are all first-generation antihistamines, just like Neoantergen and Benadryl. Although their structures varied only slightly, they have one common side effect. Although all are efficacious in treating allergies, they all cross the BBB and cause drowsiness and other CNS-related side effects.

Second-generation antihistamines are newer drugs that have fewer side effects because they do not cross the BBB in significant amounts.

§4.3. Second-Generation Antihistamines

Blocking allergies without causing drowsiness, nonsedating antihistamines are clearly more desirable drugs in comparison to the first-generation antihistamines. The second-generation drugs have been chemically engineered to stay out of the brain. As a result, they cause little or no sedation.

Albert Carr, Seldane, and Allegra

The first second-generation antihistamine was **terfenadine** (Seldane). When it was introduced in the United States in 1985 by Hoechst Marion Roussel, Seldane was the best-selling prescription drug of the year. Like all other antihistamines, Seldane eases sneezing and itching by blocking histamine's attack on the mucous membranes. But unlike the other antihistamines, Seldane barely passes into the brain, which means that its sedative effect is minimal. The new allergy treatment generated much enthusiasm because before then, many in the allergy field seriously doubted if such a wonder drug existed.

Seldane was discovered by Albert Anthony Carr of Hoechst Marion Roussel, now part of Aventis Pharmaceuticals.

Albert Carr was born in the 1930s in Covington, Kentucky. He went to Xavier University, a Catholic Jesuit liberal arts school in Cincinnati, Ohio, receiving a B.S. in 1953 and an M.S. in 1955. He went on to study organic chemistry at the University of Florida at Gainesville, earning a Ph.D. in 1958. That year, Carr joined Wm. S. Merrell Co. in Cincinnati, a storied company with a distinction of being associated with not one but two medical scandals in the 1960s. In 1960, Merrell Co. received FDA approval to market a cholesterol-lowering drug, triparanol, with the trade name MER/29 by supplying falsified data associated with safety studies using monkeys. MER/29 caused side effects ranging from hair loss and skin disease to eye damage in hundreds of patients. Also in 1960, Richardson-Merrell was the American marketing partner for the most notorious drug, thalidomide, which was used to treat morning sickness but caused thousands of birth defects in both Europe and Japan due to its strong teratogenicity. Thanks to the diligence and steadfastness of the FDA's Dr. Frances Kelsey, the United States was largely spared of one of the biggest medical disasters in history.

Carr discovered terfenadine in the early 1970s. But Hoechst Marion Roussel did not reveal its remarkable nonsedating pharmacology until 1973 at the American Society for Pharmacology and Experimental Therapeutics. At that time, the popular view was that a nonsedating antihistamine was unobtainable, and perhaps even a contradiction in terms. Curiously enough, Hoechst Marion Roussel did not follow up their 1973 disclosure with further publications until 1977, which fueled the naysayers' suspicions even more. Eventually, in 1978, a slew of important pharmacology reports appeared, confirming terfenadine's efficacy in treating allergies without CNS side effects. The FDA's approval of terfenadine (trade name Seldane) for marketing in 1985 solidly established that nonsedating antihistamines indeed were obtainable.

Seldane was one of the first drugs that was marketed through direct-to-consumer commercials on TV. Before Seldane, medicines were rarely advertised directly to consumers; drug companies promoted drugs to physicians instead. No new allergy treatment generated as much enthusiasm at the time as Seldane. When it was introduced in the United States in 1985, Seldane was the best-selling prescription drug of the year. Being the first nonsedating allergy drug, Seldane cornered the market, reaching $800 million at its peak. Although Seldane itself was not a blockbuster per se, the way it was marketed heralded the era of blockbuster drugs. Seldane lost patent protection in 1994.

One drawback of Seldane is that it is associated with dangerous irregularities in heartbeat among some patients taking it. Because Seldane is metabolized by mostly CYP450 3A4, a subtype of cytochrome 450 in the liver, it has serious drug-drug interactions with other drugs such as ketoconazole or erythromycin that use the same CYP subtype for metabolism. In addition, because Seldane causes QTc prolongation and thus causes an irregular heartbeat, its cardiotoxicities are of serious concern. In 1992, the FDA ordered the manufacturer Hoechst Marion Roussel to add a warning label to the drug.

Knowledge is power. Extensive pharmacokinetics investigation saved the day. It was known that terfenadine (Seldane) was metabolized by CYP450 3A4. In fact, terfenadine (Seldane) is completely oxidized by this particular enzyme in the liver. One of three methyl groups on the *tert*-butyl group at the end of the terfenadine molecule is totally oxidized to the corresponding carboxylic acid. That compound later became **fexofenadine** (Allegra). In other words, fexofenadine is the active metabolite of terfenadine, and terfenadine may be viewed as fexofenadine's prodrug. Fexofenadine has negligible hepatic metabolism in humans and is recovered mainly in an unchanged form after oral administration. Not only is fexofenadine as good a nonsedating antihistamine, it does not have the QTc prolongation and the drug-drug interaction issues that plagued terfenadine. The FDA proactively worked with the manufacturer closely to transition patients from Seldane to Allegra. This was apparently a win-win situation for both the regulators and the drugmaker.

Albert A. Carr received the 1999 Perkin Medal in recognition of his creativity and innovation in the discovery of new therapeutic agents.[10]

Schering-Plough's Claritin and Clarinex

By 1977, a report was published by Hoechst Marion Roussel on terfenadine (Seldane)'s efficacy in inhibiting histamine-induced skin wheals in humans and its lack of CNS side effects. That report opened a floodgate of research activities in the field of nonsedating antihistamines. As soon as Schering-Plough realized that nonsedating antihistamines in general and Seldane in particular were for real and doable, they scrambled to catch up in order to protect their significant franchise of allergy drugs.

Schering-Plough's chemists took two approaches. One was to reduce the lipophilicity (logP, a measure of the drug's greasiness) of selected existing antihistamines in order to reduce brain penetration. The rationale was that molecules needed to be reasonably greasy to cross the BBB. A significantly less lipophilic molecule would be less likely to go into the brain. The other approach was to systemically modify the structure of azatadine (Optimine), the most potent antihistamine available to them, to reduce its CNS effects. As we will see later, the two approaches would eventually coalesce to result in nonsedating antihistamines that would be synonymous with Schering-Plough.[11]

Medicinal chemist Frank Villani, the inventor of Chlor-Trimeton and Optimine, had a fascination with tricyclic compounds. As a matter of fact, tricyclic antidepressants were one of the very few choices for patients suffering from depression before the emergence of SSRIs at the end of the 1980s. It was not that tricyclics didn't work; they worked alright. The problem was toxicity. Tricyclic antidepressants were "dirty" drugs—hitting so many biological targets that they had severe side effects, such as flushing, sweating, and constipation. Only a small percentage of patients were able to finish the full course of treatment. Today, tricyclic antidepressants are very rarely prescribed. Instead, SSRIs like Prozac, Zoloft, Paxil, and so on are the gold standard for treating depression. Because of their more favorable safety profiles, they are much better tolerated, and there are fewer side effects.

Villani studied tricyclic antipsychotics like promethazine (trade name Phenergan) and amitriptyline (Endep). Inspired by these molecules, he made a rigid version of Chlor-Trimeton, an earlier Schering antihistamine. This modification led to his discovery of azatadine (Optimine), which was introduced in 1973. Villani continued his research with tricyclic antihistamines, hoping to find nonsedating drugs. However, his efforts were not looked upon kindly by the management, which eventually shut down the program. It was rumored that the head of research at the time flatly asserted that there were no such things as nonsedating antihistamines. Despite several and increasingly testy warnings to move on to other projects, Villani doggedly kept pursuing this class of tricyclic compounds. His behavior might not have engendered too much affection from the top, and Villani paid for it at the end of his career.

After Hoechst Marion Roussel's success of terfenadine (Seldane), Villani tried to attach the tail of terfenadine to Schering's pheniramine (Trimeton). Unfortunately, the chimera compound was too weak, so that approach was quickly abandoned. Instead of going back to the drawing board, Villani went back to his old love, tricyclic compounds. Simply adding an ethyl carbamate yielded a compound with good efficacy and safety. The only shortcoming of the compound was its short half-life. Pulling the old trick that he himself used for creating Chlor-Trimeton, he installed a chlorine atom and solved the bioavailability problem. The compound would later become loratadine, which was discovered by Villani and his associate Charles V. Magatti at the end of the 1970s. A patent was filed in June 1980 for a group of chemical compounds that included loratadine, which would eventually be known as Claritin. The Schering patent, issued on

August 4, 1981, stated that the compounds, including the future Claritin, were "useful as antihistamines with little or no sedative effects."

In the 1980s, it typically cost a pharmaceutical company roughly $500 million to develop a new drug, a figure that also accounted for the failures along the way. Today, the cost is more than $1 billion to develop a drug. After preclinical animal tests proved satisfactory, Schering-Plough began to test loratadine in human patients, first to establish its safety in Phase I and then to prove its effectiveness in Phase II.

One small change caused a big delay. Instead of capsules, Schering-Plough decided to switch the formulation to tablets. The switch alone took more than two years for them to show that the capsules used in clinical trials were pharmacologically identical to the tablets they intended to sell. "Bioequivalence" was what the FDA was looking for with regard to the two different formulations.

Satisfied with the results from both Phase I and Phase II clinical trials, the company recruited thousands of human patients for the pivotal Phase III trials, which were, and still are, the most expensive stage of developing a drug. They applied the gold standard in drug testing: randomized, double-blind trials pitting loratadine against both placebo and one of several already approved antihistamines. Phase III trials soundly proved that loratadine was both safe and efficacious for all the doses tested, including the lowest 10-mg dose, whose efficacy was controversial among some experts in the field at the time.

The fate of a drug could occasionally be affected by human factors.[12] In September of 1989, an officer at a division of the FDA in charge of reviewing loratadine concluded that all outstanding issues regarding safety, efficacy and the bioequivalence of tablets and capsules had been resolved. Loratadine was poised for approval at that point. Alas, due to an internal reorganization at the FDA, the loratadine application was transferred to a different division within the FDA with a new pharmacology reviewer. This reviewer revisited some toxicology issues that Schering thought had already been settled; this caused further delay of the approval.

In 1989, high doses of doxylamine (Unisom), an antihistamine and sleep aid, produced liver tumors in rodents. Meanwhile, a patient concurrently taking Seldane for a sinus condition and the antifungal drug ketoconazole for a vaginal yeast infection experienced a severe drug-drug interaction with potentially fatal cardiac arrhythmias.

At the end of 1989, the demises of two competitor antihistamines, doxylamine and Seldane, turned out to be a boon to loratadine's fortunes. Suddenly, it became the only viable antihistamine in the FDA's pipeline for approval. Eventually, after a 77-month odyssey, loratadine was approved by the FDA in 1993. Today, the deadline for completing a CDER (Center for Drug Evaluation and Research at the FDA) review for a drug is 10 months, with drug companies paying portions of the salaries of reviewers. Schering-Plough marketed loratadine with the trade name of Claritin, today a household name.

To make up for the lost time when loratadine was waiting for FDA approval, Schering-Plough's marketing machine went to work in full drive. In 1997, the FDA relaxed its rules for TV commercials for drugs. Drug companies were now allowed to prompt the viewer to "ask your doctor" for more information about the drugs. Schering-Plough poured $322 million into pitching Claritin to consumers in 1998 and 1999, far more than for any other brand. The aggressive marketing paid off. Claritin had annual sales of $1.4 billion in 1997 and jumped to $2.6 billion by 2000. It accounted for nearly 30 percent of Schering-Plough's annual revenues. It was one of the biggest blockbuster drugs of the time.

As mentioned before, Allegra is the active metabolite of Seldane, but Allegra is a much better drug with fewer side effects. Similarly, Schering-Plough's follow-up drug, the nonsedating antihistamine Clarinex, is the active metabolite drug of Claritin. When a patient takes Claritin, the liver cleaves the most vulnerable bond, which is an ethyl N-carbamate group, giving the NH group. The only difference from Allegra is that Clarinex did not seem to be that much more superior to Claritin. Nonetheless, Clarinex, marketed since 2002, provided Schering-Plough with new intellectual properties and revenues that filled the gap left by the patent expiration of Claritin.

Claritin and Clarinex, both blockbuster drugs, made Schering-Plough one of the biggest pharmaceutical companies in the world. As for the discoverer Villani, rumor has it that he was eased out the door (not so "eased" according to some), a fate not uncommon for many who discovered drugs. Sadly, one hears many stories like that these days.

5

Blood Thinners

From Heparin to Plavix

> *Why does my blood thus muster to my heart,*
> *Making both it unable for itself,*
> *And dispossessing all my other parts*
> *Of necessary fitness?*
>
> WILLIAM SHAKESPEARE

Three types of blood cells (figure 5.1) exist in the human body: red blood cells, white blood cells, and platelets, in addition to plasma, which takes up 55 percent of the blood's volume. Red blood cells take up approximately 45 percent of the blood's volume. They transport oxygen from the lungs to other body parts. White cells defend us against bacterial and viral invasions. Platelets (less than 1 percent of the blood), the third type of blood cells, are sticky little cell fragments that are involved in helping the blood clot, a process known as coagulation. Without platelets (even though they constitute less than 1 percent of blood), our blood would not be able to clot, and we would have uncontrolled bleeding.

However, formation of blood clots is a double-edged sword. Clots are beneficial because they heal cuts and wounds; blood clots in the bloodstream are harmful because they block coronary arteries, constrict vital oxygen supplies, and cause heart attacks and strokes, more and more frequent modern maladies as the baby boomers get older.

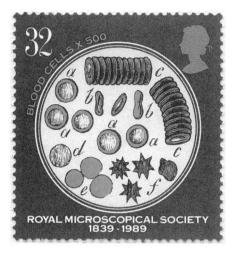

FIGURE 5.1 Blood cells. © Royal Mail.

Whenever the body is cut or injured and blood comes into contact with cells outside the bloodstream, a tissue factor on these cells encounters a particular protein within the blood, which triggers the clotting process. In the same vein, a series of other blood factors then come into action and amplify one another to quickly form a jelly-like blood clot. Blood clots form when an enzyme called thrombin marshals fibrin (a blood protein) and platelets (tiny cells that circulate in the blood) to coagulate at the site of an injury. Individuals with no ability to clot have a genetic condition called hemophilia; such people are also known as "bleeders." Queen Victoria was hemophilic, and she passed on her genes to her many heirs who ruled Europe for over a century. This is why hemophilia is sometimes known as the royal disease. Symptoms of hemophilia manifest only in male offspring.

People with hemophilia must periodically administer a clotting factor to their blood to prevent constant bleeding. But too much clotting factor in the blood makes it thick and can cause stroke and other cardiovascular diseases. For instance, coronary thrombosis is a life-threatening blood clot in the artery. Clots kill some 200,000 hospital patients in the United States each year. Deep-vein thrombosis (DVT) is commonly associated with long-haul air travel, where passengers are confined to cramped spaces for many hours. Former Vice Presidents Dan Quayle and Dick Cheney both made headlines when they suffered DVT linked to air travel.

Different from thrombosis where the clot is stationary, an embolus occurs when a clot migrates from one part of the body through the circulatory system and causes blockage. Pulmonary embolism takes place when emboli travel to the lungs. The word "embolism" was coined in 1848 by Rudolph Carl Virchow (figure 5.2), a German doctor known today as the "father of pathology."

In order to prevent and treat both thrombosis and embolism, blood thinners are the drugs of choice. Traditionally, heparin and warfarin have been widely used as blood thinners to prevent the formation of blood clots for over half a century.

FIGURE 5.2 Rudolph Carl Virchow. © Deutsche Post.

§5.1. Tales of Two Old Drugs: Heparin and Warfarin

The Discovery of Heparin

Along with aspirin, insulin, and warfarin, heparin is one of the oldest medicines still in widespread clinical use. Large-dose heparin injections are routinely used to prevent blood clotting in patients undergoing kidney dialysis or heart surgery.

Heparin was discovered in 1916 at Johns Hopkins University. Its discovery was one of the most acrimonious controversies in medical history, and numerous articles have been written on the subject. The collective consensus is that Jay McLean discovered the anticoagulant from dog liver in the laboratories of William Howell in 1917. Two years later, Howell and his student L. Emmett Holt Jr. further purified the anticoagulation substance and christened it "heparin."

William Henry Howell (1860–1945) was a giant in American physiology.[1] A Baltimore native, he attended Baltimore City College and subsequently entered the newly founded Johns Hopkins University to study medicine in 1879. His Ph.D. thesis "The Origin of the Fibrin Formed in Coagulation of the Blood" was focused on blood coagulation, which became his lifelong dedication. After receiving his doctorate in 1884, Howell taught at the University of Michigan and Harvard University for some years but ended up returning to Johns Hopkins where he stayed until his retirement. At age 36, he had already published two highly influential textbooks: *An American Textbook of Physiology* and *Textbook of Physiology for Medical Students and Physicians*. He served as the president of the American Physiological Society for 12 years beginning in 1905 when he was only 39. By 1910, Howell's research was almost exclusively focused on blood coagulation. In 1916, Howell took on a second-year medical student to work in his laboratories. The student's name was Jay McLean.

McLean (1890–1957) had a tough life.[2] He lost his physician father at age four, and his mother remarried when he was nine. The great 1906 earthquake and fire in San Francisco destroyed their house and his stepfather's business when he was 15. After two years at the University of California at Berkeley, McLean had to leave because of lack of funds, and he worked in a goldmine for 18 months to earn money to finance his education. Even after returning to school, he had to do various odd jobs to supplement his meager savings. He graduated in May 1914, having also completed first-year medical school requirements except organic chemistry, and was completely broke. But all the hardship fueled his determination to accomplish something by his own abilities.

Because McLean had studied physiology using Howell's textbook, he was extremely motivated to study medicine with Howell at Johns Hopkins University to fulfill his ambition to become a physician. After drilling oil wells for another 15 months (he had found that manual labor paid much more than white-collar work), he saved enough money and traveled to Baltimore in September 1915, even though Johns Hopkins had already notified him that he was not accepted as a student. Luckily for him, and for medicine, there was an unexpected vacancy, and he was admitted to the medical school as a second-year student. He immediately approached Howell and asked to work in his laboratory without pay for one year before his savings ran out. Howell gave him the problem of determining the value of the blood-clotting substance that Howell's group had isolated from dog brain. McLean worked nights and weekends, alone. He later chose to work with heart and liver to isolate the blood-clotting substance after reading a German article. Using Howell's method, McLean isolated a brown, waxy, and "fishy" powder from dog liver using ether and alcohol. True enough, many substances that he isolated had blood-clotting properties. Surprisingly, a couple of substances that he carefully saved for some time (which deteriorated and decomposed in the course of weeks) lost their clotting ability. Instead, they behaved like an anticoagulant—they prevented blood from clotting!

When McLean told the boss of his observations, Howell was incredulous, suspecting that the substance might be contaminated with salt, which has some anticlotting properties. McLean added his substance to a beaker with some cat blood, placed it on Howell's laboratory table, and asked Howell to notify him when it clotted.

It never did. McLean published an article titled "The Thromboplastic Action of Cephalin" in the *American Journal of Physiology* as the sole author.[3] Although Howell initiated and directed the project, he declined to join as a coauthor, possibly because he felt that the data were still premature and would have preferred to have more experiments done. To be fair, McLean did not himself fully appreciate the importance of his discovery. He left Baltimore in the summer of 1916 and did not return to this topic until years later when its significance became clear.[4]

Inspired by McLean's observation that his crude liver extract contained a powerful anticoagulant factor, Howell pursued the matter with rigor. In 1917, Howell, along with his student Holt, further purified that substance from dog liver and christened the material "heparin" (*hepar* is Greek for liver). Although heparin was later found to exist in many other organs, the name heparin stuck. In his paper with Holt and in his lectures, Howell always clearly and graciously acknowledged McLean's initial discovery, and McLean in turn always acknowledged Howell's grace. However, as time passed and Howell's contributions mounted, the majority of researchers presumed that Howell was the sole discoverer of heparin. Resenting being marginalized, McLean launched a clandestine campaign to claim the credit due to him while Howell was still alive, a campaign that became full-blown and more public after Howell passed away. Although his whole life had been rife with hardship and difficulties, McLean was successful with regard to the saga of heparin. Eight years after his death in 1959, the *Journal of the American Medical Association* (*JAMA*) published an editorial titled "Jay McLean (1890–1957), Discoverer of Heparin,"[5] a resounding validation of his contributions.

One of the major forces behind McLean's cause was Charles H. Best of insulin fame. Along with Frederick G. Banting (figure 5.3), Best codiscovered insulin at the University of Toronto in the summer of 1921. But only Banting and the lab head J. J. K. Macleod received the Nobel Prize in 1923 for this monumental discovery. Denied the laurels he rightfully deserved, Best could easily identify with the "little guy" whose discovery was misappropriated by the lab boss. Indeed, Best became McLean's staunch ally and vigorously championed his claim.[6] More importantly to science, under Best's leadership, Toronto scientists made great strides in the field of heparin and contributed decisively to

FIGURE 5.3 Charles H. Best and Frederick G. Banting. © Canada Post.

heparin's clinical applications by purifying it and providing the international standards for its potency.

Heparin for Humans

Today, we know that even the "purified" substance that Howell and Holt prepared contained only 1–2 percent of heparin.[7,8] One could only imagine how much heparin actually existed in McLean's early preparations. At the time, Howell used his influence and convinced the Baltimore firm Hynson, Westcott and Dunning to commercially produce heparin from dog liver, but their heparin never exceeded the potency achieved by Howell and Holt. The material contained impurities that caused allergic reactions and would not be usable in humans as a viable anticoagulant. Howell actually worried constantly that Hynson, Westcott and Dunning would discontinue its heparin production because the company did not make any profit from the complicated process. More fruitful investigations were carried out in both Toronto and Stockholm to get heparin pure enough to be used in clinics.

Charles Best returned to the University of Toronto in 1929 from his doctorate training in England under Nobel Laureate Henry Hale. Because of his experience in isolating insulin from dog pancreas, Best visualized a similar advance in purifying heparin. Enlisting assistance from two organic chemists, Arthur Charles and David Scott, they embarked on the pursuit with full steam. Although Howell at the time had realized that he could have moved heparin ahead faster with the help of organic chemists, he never did, choosing to go it alone by himself instead.

Charles and Scott had tremendous success in purifying heparin between 1933 and 1936.[9] They found that cheap beef lung contained as much heparin as expensive liver. Furthermore, they also discovered that if they let the liver or lung tissues decompose longer, they got much higher yields of heparin. Because heparin in tissues was bound to proteins, they used the enzyme *trypsin* to destroy the protein-heparin complex first and obtained even purer heparin. But the crown jewel of their achievements was the preparation of crystalline heparin as its barium salt. The crystalline heparin rendered a uniformly consistent composition and potency. Because heparin's barium salt was not amenable to large-scale production, they later converted the crystalline heparin barium salt to heparin sodium salt, the form that is still in use today. Thanks to crystalline heparin, which was 28 times more potent than the ones made by Howell, the University of Toronto was in a position to establish the international standard for heparin's biological activity. Best was very proud to add another feather to his cap, though the University of Toronto, of course, was also responsible for making the international biological standard for insulin. In 1935, Best and Gordon Murray, a surgeon in Toronto, carried out clinical trials with purified heparin, a significant milestone in the history of heparin.

Finnish chemist Erik Jorpes (born on Kökar, Åland, in 1894) further extended the chemistry of the Toronto heparin, which was in fact different from that found by McLean and Howell.[10] He identified it as an acidic sulfated polysaccharide. At the time, the pioneer Swedish cardiac surgeon Clarence Crafoord had already operated with success on two patients with massive pulmonary embolisms. Using material supplied by Jorpes, he initiated a research program in 1935 using heparin to prevent postoperative thrombosis and embolism. The Finnish stamp (figure 5.4) commemorates the purification of heparin by Jorpes and shows the antithrombin-binding polysaccharide sequence in the heparin molecule.[11]

These days, nearly a century after its discovery, heparin is still used extensively in clinics and hospitals for hemodialysis,

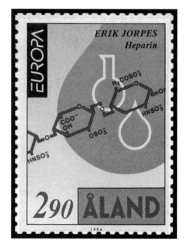

FIGURE 5.4 Erik Jorpes, heparin. © Finland Post.

vascular surgery, and organ transplantation. Open-heart surgery would not be possible without heparin. It is no exaggeration to say that heparin has helped to save millions of lives. Among them, Presidents Nixon and George Bush Sr. were treated with heparin for their heart problems. When former Vice President Dan Quayle had a blood clot lodged in his lung (pulmonary embolism), he was put on heparin as well. Another popular application of heparin is in stents. A stent is a flexible metal-mesh cylinder that acts as a scaffolding to prevent an artery from collapsing after an obstruction has been cleared in a procedure called angioplasty. A stent coated with heparin eliminates the need for a patient to take injections of heparin or other anticlotting drugs. Heparin prevents the formation of new blood clots that could block the artery at the site of the cleared obstruction.

We now also have a better understanding of heparin's mechanism of action (MOA). Heparin binds to the active site on the surface of the plasma protein antithrombin, converting this "sleeping" serine protease inhibitor antithrombin III (also known as AT III) into a potent anticoagulant. Of course, today's heparin treatment is much more sophisticated in comparison to heparin use in the past. For instance, because natural heparin, known as unfractionated heparin, has a long range of molecular weights from 3,000 to 15,000, there is always a risk of bleeding when large heparins stay in the body longer than necessary. As soon as the connection between heparin and AT III was established, it became clear that a specific active site responsible for the anticoagulation activity was present only in part of the molecules. Therefore, heparin could be fractionated using AT III activity as guidance. Modern heparins in use are mostly small molecular-weight heparins and pentasaccharides. Small molecular-weight heparins are fragments of unfractionated heparin produced by controlled enzymatic or chemical depolymerization processes that yield polysaccharide chains with a mean molecular weight of about 5,000. Pentasaccharides, on the market since 2001, are synthetic drugs that contain only the five key sugar units. One of the commercial pentasaccharides, fondaparinux, invented by Jean Choay (absorbed by Sanofi in 1983) in France, has a half-life of 17 hours; the other pentasaccharide idraparinux, the hypermethylated fondaparinux, has a half-life of 80 hours, the longest-acting pentasaccharide.

Tainted Heparin

In 2008, heparin was at the heart of a medical scandal, underscoring the level of sophistication of high-tech criminals. Tainted heparin supplies from China were linked to more than 80 deaths and hundreds of illnesses in the United States.[12] (Worldwide, the death toll was 149.) Heparin is frequently prepared from the intestinal mucous membranes of slaughtered pigs, which are often cooked in unregulated workshops. Although China has five times the pig population of the United States, a virus plague in 2007 decimated its hog population.

Because sulfur content is distinctive for heparin, many government agencies tested the sulfur ratio as a surrogate for purity. When porcine intestines were in short supply and heparin orders were hard to fill, some individuals with chemistry knowledge added 5 to 20 percent of oversulfated chondroitin sulfate, a cheaper substance made from animal cartilage, to mimic heparin. Oversulfated chondroitin sulfate is a virtual mimic of heparin in most tests and costs only one hundredth of crude heparin. This was actually a cruel twist of history. In the 1930s, Erik Jorpes was able to solve many of heparin's chemical mysteries because he had worked on chondroitin sulfate in the 1920s. His intimate knowledge of chondroitin sulfate greatly aided his investigations of heparin.

The tainted heparin scandal is reminiscent of one of the worst food-safety scandals in decades involving melamine. The amount of protein present in milk is commonly determined by analysis of total nitrogen by the Kjeldahl method, discovered by Dutch chemist John Kjeldahl in 1883. Greed drove some crooks to add urea to artificially inflate the protein content for inferior products. However, urea is easy to detect. So, some high-tech criminals came up with melamine, a fire retardant that is cheap, readily available, and has a whopping nitrogen content of 66.6 percent. In early 2007, melamine-laced white gluten in animal food imported from China sickened thousands of pets. In 2008, melamine-tainted dairy products killed six children and sickened over 300,000 in China.[13] It turns out that melamine recrystallizes in kidneys, and nearly 53,000 small children in China developed kidney stones, some as large as grapes.

Sweet Clover Disease and Dead Cows

Needless to say, heparin is of vital importance in preventing blood clotting during dialysis and surgeries. However, it has to be given intravenously, and an oral blood thinner would be much more convenient for the patient who has to take it for the long term.

Warfarin fills such a void. While the credit for the discovery of heparin was contentious, no one disputes the fact that warfarin was discovered by Karl Paul Link. Whereas heparin was isolated from animal organs, warfarin traces its roots to plants.

In the winter months of 1921–22, on the prairies of North Dakota and Alberta, Canada, cattle farmers saw their livestock die in droves. Some apparently healthy animals were dead within a few hours after dehorning. Others died from internal hemorrhaging. Frank W. Schofield, a pathologist at Ontario Veterinary College, established the link between the deaths and spoiled sweet clovers.[14] Sweet clovers (*Melilotus alba* and *Melilotus officinalis*), imported from Europe in the early 20th century, proved to be nutritious for cows and sheep when used as fresh fodder or hay.[15] During the winter, some hay became moldy and would normally have been discarded. But it was the Great Depression, and poor farmers who had only spoiled hay to feed their stock used it and saw their animals plagued by "sweet clover disease." Methodically, Schofield carried out a series of well-planned experiments using rabbits as the animal model. He solidly made the connection that it was the moldy hay that caused the excessive bleeding and killed the cattle. Regrettably, the college administration did not allow him to continue his research and ordered him back to teaching; Schofield was so disenchanted that he went to Korea as a missionary instead. By 1931, Lee M. Roderick of the North Dakota Agricultural Experimental Station in Fargo, Minnesota, found that sweet clover disease could be controlled by withdrawal of spoiled hay or transfusion of fresh blood from healthy cattle.

What happened next became a legend in medical folklore. While a blizzard was howling and the mercury was below zero on a Saturday in February 1933, Ed Carlson, a farmer from Deer Park, Wisconsin, drove 190 miles to Madison.[16] He had already lost half a dozen cows in the last two months, and his local veterinarian could not do anything about it. Carlson drove to the Agricultural Experimental Station in Madison, bringing a dead heifer, a milk can containing blood completely devoid of clotting capacity, and about 100 pounds of spoiled sweet clover. But the station was closed; pure chance brought him to the biochemistry building where agricultural chemist Karl Link was still at work. After hearing about Carlson's predicament, Link told him what Schofield and Roderick found and advised him that he should feed the cows only fresh hay or give them blood transfusions. That was not really a satisfactory solution for the poor farmer, as Link's colleague Eugen Wilhelm Schoffel commented in his Swabian German accent: "Get some good hay...transfuse. Ach!! Gott, how can you do dat ven you haf no money?" That chance encounter would compel Link to redirect his research to find a cure for sweet clover disease, although he had been mostly working on carbohydrate chemistry before Carlson showed up that day.

Link (1901–1978) was born in the Lutheran Parsonage at La Porte, Indiana, as the eighth of ten children.[14] He earned his Ph.D. in agricultural chemistry at the University of Wisconsin in 1925 and studied in Europe under Nobel Laureates Fritz Pregl and Paul Karrer. Returning to Madison in 1927, Link was first appointed as an assistant professor but was quickly promoted to be the first professor of biochemistry at the University of Wisconsin. After his encounter with the hapless Carlson, Link spent the next six years trying to isolate the anticoagulant from spoiled sweet clover hay without much success. His long and arduous journey ended on June 26, 1939. When he arrived at the laboratory that morning, his assistant Chet Boyles was drinking, claiming: "I'm celebrating; Doc. Campy has hit the jackpot."[16]

Although Mr. Boyles had always been a drinker, he had a genuine reason to celebrate on that particular morning. Link's associate Harold A. Campbell had isolated 6 mg of crystals of the principal hemorrhagic agent. The task of deciphering the structure of the compound fell to the sensitive, brilliant, and deft chemist C. F. Huebner in Link's group, who with some assistance from his lively imagination made the correct structural elucidation. He then immediately set out to synthesize the same compound, later named dicumarol, and accomplished the feat on April fool's day, 1940. Dr. Huebner was nobody's fool; the compound he prepared behaved identically in every way to the naturally isolated dicumarol from the moldy hay. Now we know that dicumarol was formed when coumarins in the sweet clover were oxidized during fermentation with mold.

Warfarin, the Rat Poison

Within two years after dicumarol was synthesized, over 100 analogues were prepared in Link's laboratory. Clinicians were initially enthusiastic about using dicumarol as an anticoagulant for humans, but it never became a drug.

Early in September 1945, shortly after V-J Day, Link took a trip with his family and caught pulmonary tuberculosis. He was sent to a sanatorium to recuperate for six months. Although he did not complain about the treatment, which entailed three beers a day, he was bored and took up reading about the history of rodent control. Back at the lab, Link announced that he propose dicumarol analogues for rodenticidal use. The news shook everybody up, and the consensus was that "the boss has really gone off the deep end this time."[16] With the help of the good old Schoffel, Link explored dicumarol as a rat poison. But they found that dicumarol was not a good choice because it acted too slowly. But number 42 of dicumarol's analogue (WARF-42, WARF stands for Wisconsin Alumni Research Foundation) proved ideal in many respects. Little did they know that it would become the most successful rodenticide *and* the most used oral anticoagulant in history.

In 1948, Link approached his research sponsor WARF and asked them to patent WARF-42 as a rodenticide with the name warfarin (trade name Coumadin). Link came up with the seemingly belligerent name warfarin by combining WARF with coumarin. A truly loyal U. Wisconsin man, he turned over all royalties from his patents to WARF. Just as cattle ate hemorrhagic sweet clover hay until they died without sensory response, the rat ate cereal bait laced with warfarin until fatal hemorrhage set in. Water-soluble warfarin made it possible for the rat to drink itself to death. Warfarin was the safest rodenticide known at the time, and it became widely used with no recorded case of warfarin-induced fatality in humans.[16]

Warfarin, the Blood Thinner

Initially, clinicians had no qualms about testing dicumarol on humans; but there was little interest in testing warfarin, the rodenticide, as a blood thinner. Who could blame them? The drug used to exterminate rats and mice was surely too toxic for humans. However, a report from Captain J. Love of the U.S. Navy Medical Corps in Philadelphia radically shattered that notion.

In early 1951, an Army inductee attempted to commit suicide by swallowing rat bait containing 567 mg warfarin in cornstarch for five days, every day. Even though he followed the multiple-dose directions on the package, it became apparent that he simply could not kill himself. With enough time to have second thoughts, he walked to the base clinic with a fully developed case of hemorrhagic sweet clover disease. He was treated with a blood transfusion and a large dose of vitamin K and immediately recovered.[16]

That incident became a catalyst in transforming warfarin from a lethal poison for rats to a blood thinner for humans.[17-21] Warfarin is more potent than dicumarol. It also has higher water solubility and bioavailability in comparison to dicumarol.[17] But the best thing about warfarin and other coumarin analogues is that there is an antidote—vitamin K. As early as 1942, Link's group already knew that vitamin K could reverse the hemorrhaging of rats treated with warfarin. Vitamin K is a key player in blood clotting. Warfarin, in turn, works as a vitamin K antagonist by inhibiting the vitamin K epoxide reductase. Vitamin K was discovered by Danish scientist Henrik Dam in 1929. He called it a coagulation vitamin (*koagulation* in German), which was where the vitamin got its letter K. Dam shared the 1943 Nobel Prize in Medicine for his work on vitamin K with Edward A. Doisy of Saint Louis University, who also did much of the research that led to the discovery of the structure and chemical nature of vitamin K.

While clinical trials for warfarin as a blood thinner proceeded slowly in the early 1950s, another incident attracted national attention to warfarin as an anticoagulant.

In September 1955, Link's warfarin was given to President Eisenhower after he suffered a coronary thrombosis.[18] Although it was kept a secret, it did not take long to surmise that the most important man in the world at that time was treated with warfarin sodium as an anticoagulant.[16] In a completely unrelated story, it was rumored that Joseph Stalin died of warfarin poison. Historians found that his symptoms before his death were archetypical of warfarin poisoning. This speculation may be taking too much license with history. Yes, warfarin is tasteless and odorless, a perfect poison to blend with food or drink, but warfarin was not really widely used until after Stalin's death in 1953. Nikita Khrushchev and his comrades might not have been able to procure warfarin and learn to employ it properly. It would have taken five to ten days of large doses of warfarin to achieve the extent of massive hemorrhage in both the left brain and stomach that occurred during Stalin's death throes. As we learned from the failure of the Army recruit to commit suicide by digesting a copious amount of warfarin rat bait, warfarin is not a reliable poison for killing a human being, either quickly or slowly.

Another piece of history, certainly not fictional in this case, was President Richard M. Nixon's use of warfarin.[21] On October 27, 1974, the 61-year-old politician was hospitalized for chronic recurrent venous thrombosis and a recent pulmonary embolus after his frequent international trips. A course of warfarin was begun, and the famous patient was discharged shortly thereafter. Because of a compromised prothromin level, Nixon was readmitted to the hospital and given increasing doses of warfarin. Two days later, an operation successfully removed his thrombus. Curiously, Nixon died of a stroke[19] years later after his warfarin treatment had been, bewilderingly, stopped.

Even today, warfarin is still one of the most used oral anticoagulants. It is cheap, and there is always vitamin K as an antidote when there is an overdose. However, warfarin has been occasionally associated with side effects such as skin necrosis and hair loss. Moreover, gauging the dosage of warfarin is notoriously imprecise and tricky. It depends on factors such as a patient's age and weight. Another factor is genetic polymorphism in the genes encoding CYP2C9, the main enzyme responsible for the metabolism of S-warfarin, the more potent of the two enantiomers.[22] Additional factors include the varying natures of VKORC1 (Vitamin K epoxide reductase complex, subunit 1) and vitamin K epoxide reductase, the enzyme warfarin inhibits. Even a slight change in dosage can mean the difference between too little, which would not be effective in preventing blood clots, and too much, which can cause dangerous internal bleeding. Tens of thousands of people end up in the hospital each year with complications from warfarin.

Since genetic polymorphisms in CYP2C9 and VKORC1 are important factors for determining warfarin's dosage, several diagnostic companies developed warfarin response tests, which cost $50 to $500, to look at variations in these two specific genes in each patient. Such tests might help tell which drug would be best for a particular person, or whether a patient might be susceptible to dangerous side effects. Regrettably, the Centers for Medicare and Medicaid Services have decided not to pay for the genetic tests, arguing that there are too many other factors that might influence warfarin's optimal dosage.[23]

§5.2. Good Old Aspirin

In the 1890s, when the German drug firm Bayer began to market aspirin, the management's biggest concern was its potential harmful cardiovascular effects. Today, aspirin is promoted as a popular medicine to *prevent* cardiovascular diseases. What a difference a century of scientific advancement (figure 5.5) could make!

Aspirin is prepared from salicylic acid, an ingredient found in willow trees and meadowsweet. The *Ebers Papyrus* referenced willow's medicinal properties in general and its treatment of rheumatism in pregnant women in particular. Ancient medicine men including Hippocrates, Pliny, Celsus, Discorides, and Galen all described the medicinal use of willow tree as an analgesic (to reduce pain) and antipyretic (to reduce fever).[24] Widespread use of salicylic acid became possible in the mid-1850s when Hermann Kolbe discovered how to make it from phenol and carbon dioxide. Unfortunately, salicylic acid tastes very bitter and can easily damage the stomach because of its strong acidity (its pKa is 3).

FIGURE 5.5 Aspirin. © Deutsche Post.

Aspirin, acetyl salicylic acid, is relatively easier on the GI tract and is now one of the most used medicines in the world.

Although Charles Gerhardt in France and Karl Johann Kraut in Germany both synthesized acetyl salicylic acid in crude forms, neither pursued its medical applications. Felix Hoffmann (figure 5.6), a chemist at Bayer and Company in Germany, is credited with making aspirin. As legend has it, Hoffmann's father

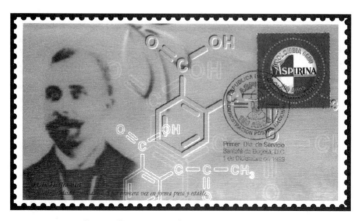

FIGURE 5.6 Felix Hoffmann. © Administracion Postal Nacional, Colombia.

had severe rheumatism and could not tolerate sodium salicylate and begged him for a better drug. Tinkering with salicylic acid analogues and perhaps inspired by Gerhardt and Kraut's chemistry, Hoffmann prepared acetyl salicylic acid (which later became aspirin) on October 10, 1897. Hoffmann also made acetylated morphine, which later became heroin. Ironically, while Bayer management was more enthusiastic in promoting heroin, they were initially reluctant to market aspirin because salicylates were believed to have detrimental effects on the heart. Frustrated by his superiors' inaction, the head of Bayer's pharmaceutical section, Arthur Eichengrün, took aspirin himself and did not find any problems with his heart. After that, he began distributing free samples to physicians at conferences. Quickly, aspirin became a popular drug in Germany, then in Europe and the whole world.

Coincidentally, Karl Link, the father of warfarin, was one of the first to explore aspirin's antithrombotic effects. Because dicumarol is metabolized into salicylic acid, it could be viewed as a prodrug of salicylic acid, just like aspirin is a prodrug of salicylic acid as well. In that context, Link experimented on himself and found that, as an anticoagulant, aspirin was not very potent. Indeed, warfarin is about twice as potent as aspirin as a blood thinner. Link published several reports in 1943 on aspirin's anticoagulant effect although he considered this as a dangerous side effect rather than a therapeutic one.[25]

When aspirin was first marketed, physicians feared that it might cause heart problems like salicylates did. To quash this misconception, Bayer often labeled their aspirin bottles with the statement "Aspirin Does Not Affect the Heart." Ironically, few would have dreamed that aspirin would one day be taken to prevent heart attacks and strokes. Its genesis was not from a big-shot professor in a high-power university or research institute, but a practicing physician named Lawrence Craven.[26]

In the 1940s, Craven, a family doctor in Glendale, California, noticed that his patients who took large doses of Aspergum sticks (a chewing gum imbued with aspirin) for pain after surgery had difficulty in stopping bleeding. Intrigued, he wondered if aspirin possessed anticlotting properties. Since many of his affluent middle-aged patients were overweight, he recommended they take a whopping 325 to 650 mg of aspirin as a preventative anticoagulant. Miraculously, among 1,465 healthy male subjects who took aspirin, none suffered coronary occlusion or insufficiency. He published his results in the obscure *Mississippi Valley Medical Journal* in 1953. In a footnote, Craven proudly noted that his article won the third-place prize in the 1952 Mississippi Valley Medical Society Essay Contest. Unfortunately, his clinical insight was often discredited as not being sufficiently scientifically rigorous. Ironically, Craven himself died of a myocardial infarction (heart attack) in 1951, although he practiced what he preached by religiously taking aspirin. Nonetheless, many scientists at the time began to explore the antiplatelet effect of aspirin, and massive clinical trials were carried out to gauge the statistical significance.

Harvey J. Weiss at Columbia University was one of the first to rediscover aspirin's antiplatelet effect. In 1967, spurred by Armand Quick's report that low-dose aspirin prolonged the prothrombin (clotting) time in normal subjects, Weiss gave 300 mg of aspirin to ten healthy men (six of whom were his fellow physicians). He observed that aspirin indeed inhibited platelet aggregation and adenosine diphosphate (ADP) release. Meanwhile, many researchers reported that anti-inflammatory agents, including aspirin, inhibited aggregation of platelets in several species of animals. Weiss proposed in his article for *Lancet*: "the results suggest that these agents may have antithrombotic properties."[27]

Despite aspirin's popularity in treating almost every malady known to humans (the world swallows 50 million pounds of aspirin a year), its MOA was not deciphered until 1971. John Vane found that aspirin works by inhibiting prostaglandin synthetase, explaining most of its antiplatelet, antipyretic, and anti-inflammatory properties.[28] Vane, along with Sune Bergström and Bengt Samuelsson (figure 5.7), was awarded the Nobel Prize

FIGURE 5.7 John Vane, Sune Bergström, and Bengt Samuelsson. © Micronesia Post.

in Physiology or Medicine in 1982 for his discoveries concerning prostaglandins and related biologically active substances.

By the early 1980s, accumulating scientific data overwhelmingly confirmed that aspirin does prevent stroke, myocardial infarction, and ischemia brain stroke. In 1985, Margaret Keckler, the U.S. Secretary of Health and Human Services, announced to the public that an aspirin a day helps prevent a second heart attack. Baby aspirin (81 mg) for preventing heart attack has been a mainstay in many homes' medicine cabinets since then. Puzzlingly, a 10-year study involving 40,000 women in July 2005 seemed to suggest women respond less favorably than men to aspirin's cardio-protective effect. Why? We do not know.

These days, aspirin is a billion-dollar business worldwide. It is indeed a blockbuster drug, but it is not a bona fide blockbuster drug per se because the profits are divided by thousands of generic drug companies.

§5.3. From Ugly Duckling to Beautiful Swan—From Ticlid to Plavix

The Demise of Ximelagatran

Warfarin became the first oral blood thinner for humans in the early 1950s. It took another 50 years for the second oral anticoagulant to emerge: AstraZeneca's ximelagatran (trade name Exanta). Direct thrombin inhibitors such as argatroban, leprudin, and hirudin were found after warfarin, but all of them have to be administered parenterally. Hirudin, a direct thrombin inhibitor, was isolated from leech salivary glands and was the inspiration for the emergence of ximelagatran.

Medicinal leeches have been used since two thousand years ago for bloodletting to keep the body's balance of "humors."[29]

Even today, they are still employed to reduce blood coagulation during microsurgeries involving eyelids, fingers, and ears. One medicinal leech in particular has long been applied to treat strokes and heart attacks: *Hirudo medicinalis*, or the European medicinal leech. Even a minute amount of hirudin has a strong anticoagulating effect, which explains why blood does not clot when the leech is sucking on its "victim." Since harvesting hirudin from leeches is impractical, it is currently synthesized using recombinant techniques. Investigation of the MOA revealed that hirudin binds to thrombin at the site where thrombin would ordinarily bind to activate fibrinogen. Cutting off some inactive segments of hirudin's 65 amino acids, bivalirudin with only 20 amino acids was developed for intravenous administration.

AstraZeneca at Mölndal in Sweden took on the challenge of discovering orally active direct thrombin inhibitors using a pentapeptide containing hirudin's active site as their starting point (prototype). The Mölndal site already made its fame and fortune with its proton-pump inhibitor franchise, giving birth to Prilosec and Nexium, both blockbuster drugs.

In order to achieve oral bioavailability, a drug is preferred to have a lower molecular weight. In 1982, a U.S. patent revealed that a dipeptide-based drug inhibited thrombin, albeit with low potency. Encouraged by the result, AstraZeneca arrived at their first clinical candidate inogatran, which was about 100-fold more potent than the original dipeptide reported in the U.S. patent.[30,31] In the early 1990s, AstraZeneca gained access to the three-dimensional coordinates of human α-thrombin, a great boon in appreciating how a drug might bind to the active sites of this protease. With the aid of this profound knowledge, medicinal chemists were able to better define the chemical space within the target enzyme. They arrived at their own dipeptide: melagatran. The melagatran molecule has a distinctive feature: It contains an azetidine intermediate, a nitrogen-containing four-membered ring (azetidines). The chemists quickly exhausted the world's limited and expensive supplies of the azetidine intermediate, and the program needed kilograms of the material just to proceed. Luckily, they uncovered 75 kg of it gathering dust on the shelves of a European company that had extracted the compound from 5,000 tons of sugarbeets.

Unfortunately, melagatran is highly ionic, with an oral bioavailability of less than 3 to 7 percent in humans, although its bioavailability was greater than 50 percent in dogs. Although it can only be given parenterally, just like hirudin and bivalirudin, AstraZeneca was determined to make an oral thrombin inhibitor. The task was so daunting that at one point, their lead pharmacokineticist offered to "eat his hat" if they succeeded.[30] His doubts were somewhat justified: Reaching melagatran had already taken the team nine years. It took them another two years to discover an orally active drug, ximelagatran. Two changes in the melagatran molecule did the trick. One was transforming the original carboxylic acid into a corresponding ethyl ester, a very common tactic in medicinal chemistry. Another less common tactic was transforming the original amidine, a strong base, to hydroxyamidine, a nearly neutral fragment. In essence, ximelagatran is a double prodrug of melagatran. As a consequence, the oral bioavailability was increased to 18 to 20 percent in humans, which was good enough for the drug to be given orally. We don't know whether the lead pharmacokineticist ate his hat.

Thus, ximelagatran became the first oral anticoagulant in 50 years after warfarin. Unlike warfarin, which usually takes four to five days to achieve its full anticoagulant effects, ximelagatran has a faster onset of therapy. And while warfarin needs frequent therapeutic monitoring, ximelagatran does not. Warfarin is known to interact with food or drugs, whereas food-drug and drug-drug interactions are not common for ximelagatran; this offered welcome options for the management of thrombotic disease. AstraZeneca won regulatory approval from a dozen countries for ximelagatran and marketed it with the trade name Exanta. A disadvantage of ximelagatran is that there is no antidote if acute bleeding develops, whereas warfarin can be antagonized by vitamin K and heparin by protamine sulfate. Another drawback of ximelagatran is its severe liver toxicity in a small population of patients, which was one of the reasons the FDA rejected the drug in 2004 for U.S. licensing.[32] In 2006, AstraZeneca voluntarily withdrew ximelagatran from the market after reports of liver damage during additional trials.

Jean-Pierre Maffrand of Sanofi

Generals go to war with the army they have in hand, not the army they would like to have. Drugs are often discovered this way: not because of enlightenment or ingenious insight, but because chemists know how to make them. The reason why two blood thinners, Ticlid and Plavix, came to be was because a chemist knew how to make them; he was Jean-Pierre Maffrand of Sanofi.

Unlike most big drug firms, Sanofi S.A. was a latecomer and was founded only in 1973 as a spin-off of Elf Aquitaine, France's largest state-owned oil company. One year later, Sanofi bought the small drug firm Parcor at Toulouse. Along with Parcor came a medicinal chemist named Jean-Pierre Maffrand.

Maffrand graduated in 1966 from ENSC, Montpellier, where he had studied chemical engineering. He earned his Ph.D. in organic chemistry under P. Maroni in 1968 at the Université Paul Sabatier in Toulouse. After that, Maffrand carried out his postdoctoral research as part of his cooperative military service at the University of Sherbrooke in Quebec, Canada, in the laboratories of Pierre Deslongchamps. Maffrand's Ph.D. thesis focused on stereoelectronic effects, and Deslongchamps was an expert in the field. Deslongchamps had authored a book titled *Stereoelectronic Effects in Organic Chemistry*, so he was a logical choice for Maffrand. The fact that Deslongchamps taught in Quebec, a French-speaking province, was probably an added incentive as well. With Deslongchamps, he worked on the total synthesis of (+)-ryanodol, a natural product. In 1971, he started his career in pharmaceutical research in a small company called Castaigne S.A., later Parcor, in Toulouse. Toulouse is the home base of the European aerospace industry, with the headquarters of Airbus located there. Parcor was bought in 1974 by Sanofi, but Maffrand kept publishing his work using Parcor as his affiliation even after 1979 when Sanofi finally consolidated its pharmaceutical operations including Parcor, Labaz, and Galor.

In 1974, Maffrand and his colleague F. Eloy published an article titled "Synthesis of Thienopyridines and Furopyridines of Therapeutic Interest" in the *European Journal of Medicinal Chemistry*.[33] That was the first time when Maffrand mentioned thienopyridines, the core structure of Ticlid and Plavix, in his publications. In 1976, Maffrand filed his first patent (strangely, a German one) for thienopyridine derivatives, with Parcor as its assignee. From then on, Maffrand and his colleagues patented extensively during the next decade on thienopyridines and their pharmaceutical utilities. The fruit of their labor was ticlopidine (trade name Ticlid), which was marketed in 1979.

Ticlid

As early as 1972, just a couple of years into his first job, Maffrand was assigned to lead a small research team. His supervisor asked his group to make analogues of tinoridine to find drugs with improved anti-inflammatory properties. Tinoridine, a thienopyridine, was an anti-inflammatory drug discovered by the

Japanese drug firm Yoshitomi and was marketed with the trade name Nonflamin. As James Black stated, the best way of finding a new drug is to start with an old one. Maffrand and colleagues synthesized several thienopyridines, but those compounds had no anti-inflammatory properties at all. However, not all was lost. Further scrutiny revealed that they inhibited blood platelet aggregation. Immediately realizing the important roles that platelets played in myocardial infarction and brain ischemia, the team set out to find a platelet aggregation inhibitor that was superior to aspirin. The pharmacologists devised several biological tests to expedite the screening process. The tests included the bleeding-time test, the silk-thread test, the collagen test, and the ADP test. They prepared and evaluated hundreds of similar compounds to find the ones with optimum antiplatelet aggregation.

After more than five years of synthesis and pharmacological and animal testing, the team arrived at ticlopidine. A patent was filed in 1977, and its marketing was approved in France in 1978. Sanofi chose Syntex as its American co-development and co-marketing partner for ticlopidine, which received FDA approval in November 1991. Syntex sold ticlopidine using the trade name Ticlid.

Sadly, after Ticlid reached the market, a rare but potentially fatal side effect was observed: thrombotic thrombocytopenic purpura (TTP), a blood disorder where tiny blood clots form in small arteries throughout the body, destroying red blood cells and causing anemia. The disorder can also cause fever, kidney failure, slurred speech, confusion, disorientation, and coma. TTP usually appeared within two weeks after use of the drug began and required aggressive treatment with plasma exchange, a costly procedure in which a large amount of the liquid portion of the blood is removed. This was a serious problem because in the 1990s, Ticlid was routinely used in the approximately 500,000 patients per year in the United States who underwent a percutaneous coronary intervention involving a stent.[34] The mortality rate for this rare complication exceeded 20 percent. Limiting ticlopidine therapy to two weeks after stenting did not prevent the development of TTP. Ticlid can also reduce the number of infection-fighting white blood cells to dangerously low levels in about 1 percent of users and apparently causes TTP in about 1 out of every 1,600 to 5,000 patients.[35]

To make matters worse, TTP was not the only woe that Ticlid had. A cursory skimming of the medical literature, with title after title bringing more bad news, told it all:

> Severe Diarrhea Caused by Ticlid Associated with Disorders of Small Intestine Motility
> Ticlopidine (Ticlid) and Severe Bone Marrow Aplasia
> Ticlopidine-Induced Cholestatic Jaundice
> Ticlopidine-Induced Acute Cholestatic Hepatitis

Because of its association with TTP and other side effects, when Ticlid was approved by the FDA in 1991, the agency required it to carry a "black box" warning that it could cause life-threatening blood disorders. As a consequence, sales of Ticlid were insignificant.

The Discovery of Plavix

In light of Ticlid's shortcomings, a safer drug with equal or higher potency was desired. Back in 1975, Sanofi synthesized a ticlopidine derivative by adding a methyl group at the bridging carbon to yield methyl-ticlopidine. The molecule was now chiral since a chiral center was created. Pharmacological assays and animal tests indicated that it was more potent as an antiplatelet aggregation agent than Ticlid but was even *less* well-tolerated than Ticlid. The methyl-ticlopidine was therefore a less desirable drug due to its toxicity. In 1978, Maffrand asked his associate Alain

Badorc to separate the two enantiomers of the methyl-ticlopidine racemate.[36] As luck would have it, Badorc accomplished the feat on his first trial. He treated the racemate with tartaric acid (the same acid that Pasteur used to make his monumental discovery of molecular chirality; see chapter 3) in ethanol. He obtained crystalline solids that were readily separated by recrystallization. The right-handed isomer was inactive, and the left-handed isomer was active, but it was still more toxic than Ticlid. Needless to say, development of the methyl-ticlopidine was discontinued.

Similarly, the ethyl-ticlopidine was also prepared in 1978. It was still more toxic than Ticlid. This time, Badorc was not able to make crystals and separate the two enantiomers, so he prepared them by asymmetric synthesis. Unfortunately, both isomers were less potent than Ticlid. Unsurprisingly, they too were subsequently abandoned.

In July 1980, Sanofi prepared the methyl ester substituted at the bridge carbon of the ticlopidine molecule. The methyl ester contained the enantiomer that would one day become clopidogrel (Plavix). The corresponding HCl salt was prepared and designated as PCR 4099. It was found to be both more potent and better tolerated than ticlopidine. Regrettably, the racemate caused convulsions in rats, mice, and baboons at certain high dosages. In order to find out if each enantiomer behaved differently in terms of potency and toxicity, in November 1985, Maffrand asked Alain Badorc and Dr. Daniel Frehél to separate the enantiomers of the PCR 4099 racemate.

Badorc first tried asymmetric synthesis, an approach he successfully used to make the chiral ethyl-ticlopidine. But the reaction caused racemization: The chiral center scrambled, giving a 1:1 mixture of racemate. Thus he was not able to obtain the desired product with high optical purity. In the first two months of 1986, trying his luck with recrystallization, Badorc set up a screening experiment with about 30 chiral acids in various concentrations in various solvents. Checking the test tubes daily, in mid-March, he eventually observed good crystals in only one test tube one month after his experiment was initiated. The good-quality crystals were prepared by using 2.5 equivalents of (+)-camphorsulfonic acid in acetone. That was a lucky break; when only one equivalent of (+)-camphorsulfonic acid was used, Badorc did not isolate crystals.

Things always become easy once you know how to do them. Having figured out how to make crystals, the recrystallization process became rather straightforward. Badorc was able to separate the two enantiomers with relative ease. The left-handed isomer had no significant platelet inhibition activity in any assay; but the right-handed isomer was both potent and better tolerated than Ticlid. More gratifyingly, the right-handed isomer did not cause the convulsions in animals at high dosages that both PCR 4099 and the left-handed isomer did.

By screening many pharmaceutically acceptable salt forms, Sanofi found the bisulfate salt (the salt formed with sulfuric acid) of the right-handed isomer to be the most and only ideal salt form, and the salt would eventually become the drug with the generic name clopidogrel bisulfate (Plavix).

Salt selection is an essential phase of modern drug development. An estimated half of all drug molecules used in medicine are administered as salts.[37] An appropriate salt could improve solubility, increase chemical stability, and also facilitate the manufacturing process. More important, the salt form is a crucial component of intellectual properties associated with a drug, a legitimate means to achieve extension of patent protection and smooth the progress of regulatory approval.

In March 1987, Pierre Simon, then head of Sanofi's R&D division, decided to terminate the Phase I development of PCR 4099 in favor of pushing clopidogrel forward. Maffrand was initially reluctant—the development would suffer a four-year setback. But the data were compelling enough, and clopidogrel was moved up to development, with Sanofi "taking all possible steps to reduce lost time to a minimum."[36]

That was a wise decision and one that would help millions of patients. A future blockbuster drug would be born.

Plavix, the Blockbuster Drug

By 1993, Sanofi was ready to commence the Phase III clinical trials for clopidogrel bisulfate. They decided to choose an American partner. For one thing, the United States was and still is the biggest market for medicine in the world, despite having only 5 percent of the world's population. For another thing, developing a drug was and still is an extremely expensive enterprise. It could take hundreds of millions of dollars to develop a drug. Today, it takes over 1.3 billion dollars for one drug to move from the test tube to patients, according to one statistic. A codeveloping partner could share the financial burden and the risk, especially considering that as much as 40 percent of drugs actually fail in Phase III trials. At the time, Sanofi's original partner for Ticlid, Syntex, was not doing well. It would be acquired by Hoffmann La-Roche a year later. Roche then inherited the right to Ticlid. Instead, Sanofi chose Bristol-Myers Squibb (BMS) as its co-developer and future co-marketer for clopidogrel. BMS, the number-four drug company in the world at the time, had already established itself as a leading pharmaceutical company in developing and marketing cardiovascular drugs. BMS's David Cushman and Miguel A. Ondetti discovered the first angiotensin-converting enzyme (ACE) inhibitor captopril (trade name Capoten) in 1978 for the treatment of hypertension. By 1992, Capoten was a blockbuster drug, raking in $1.66 billion in annual sales.[38] More important, BMS also proved itself as a competent co-developer and co-marketer. The company was Sankyo's American partner for Pravachol, the statin drug for lowering LDL (the "bad") cholesterol. Although Pravachol was only on the market for its first year, BMS already achieved an annual sale of $400 million. It would reach $2.3 billion in 2005.

In a 50:50 deal, both Sanofi and BMS paid for the late-stage clinical development. In contrast to Ticlid, bedeviled by bad news and toxicity, the clinical data for clopidogrel seemed to be getting better and better. But the most significant and the largest-scale clinical trial was CAPRIE (Clopidogrel versus Aspirin in Patients at Risk of Ischemic Events), a randomized, double-blind trial that enrolled 19,185 patients who had suffered a recent ischemic stroke or heart attack or who had established peripheral arterial disease.[39] CAPRIE resoundingly demonstrated that clopidogrel bisulfate provided a significant improvement in the prevention of myocardial infarction, stroke, and vascular death in patients with symptomatic atherosclerosis (ischemic stroke, myocardial infarction, or established peripheral arterial disease). Whereas aspirin prevented about a quarter of such events, clopidogrel bisulfate prevented about a third. The CAPRIE study also demonstrated that clopidogrel bisulfate had a lower risk of severe gastrointestinal hemorrhage than aspirin.

Sanofi and BMS filed the NDA (new drug application) for clopidogrel bisulfate (trade name Plavix) on April 28, 1997, and the drug was granted a priority regulatory review, a process typically reserved for important new medicines for life-threatening illnesses. On November 18, 1997, the FDA granted clearance to market Plavix for the reduction of atherosclerotic events (myocardial infarction, stroke, vascular deaths) in patients with atherosclerosis documented by recent myocardial infarction, recent stroke, or established peripheral arterial disease. The drug also won approval from the European Commission. The first European launches of clopidogrel bisulfate took place in July 1998 under the name Iscover, implying that the drug helps patients recover from ischemic stroke. While BMS marketed Plavix in the United States, Sanofi marketed it in the rest of the world.

In November 1998, Sanofi S.A. and Synthélabo S.A. merged to become Sanofi–Synthélabo in a deal valued at $10.4 billion. A week earlier, Rhone-Poulenc S.A. and Hoechst A.G. merged to become Aventis.[40]

No sooner than Plavix gained FDA approval did it quickly become widely used in angioplasty procedures to open clogged arteries in the heart and legs and to prevent strokes. Within the first year, more than three million U.S. patients. had taken Plavix because physicians realized that Plavix was more efficacious than aspirin and lacked its GI side effects. Plavix pretty much completely displaced Ticlid in the marketplace. Some doctors called Plavix a "super aspirin," and it lived up to its name. Vice President Dick Cheney took Plavix for 30 days after undergoing angioplasty in November 2000 and began taking it again after another angioplasty on March 5, 2001.[41] More than 48 million Americans use Plavix on a daily basis.

Jean-Pierre Maffrand, the chemist who was principally responsible for the discovery of both Ticlid and Plavix, was justifiably treated as a hero at Sanofi. He was director of hemobiology from 1981 to 1991 and then rose through the ranks with increased responsibilities. In September 2004, he was promoted to senior vice president, director of discovery research at Sanofi-Aventis, in charge of more than 3,000 scientists at the 17 European and American sites. Maffrand retired in January 2008. He was awarded numerous scientific awards, but the coolest was the Chevalier de la Légion d'Honneur of the French Republic.

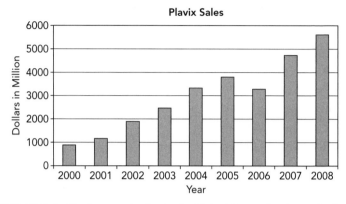

FIGURE 5.8 Plavix annual sales in the United States (by the author).

On the sales chart (figure 5.8), it is clear that annual sales of Plavix grew steadily since its U.S. launch. But there was a significant dip in 2006, which was almost $1 billion less than the natural trend would have predicted. What happened?

The phenomenon was the direct result of a blood feud between Sanofi/BMS and the Canadian generic drug company Apotex.

§5.4. Blood Feud

In 2006, Plavix was at the center of one of the most dramatic showdowns between the innovative brand-name pharmaceutical companies and generic companies trying to market copycat versions of drugs before their patent expirations.

Copycat

The world has had brand-name drug companies and generic drug companies for over a century. The old rules of engagement were that generic companies waited until patents expired and then began making and marketing the generic drugs at lower prices.

That way, brand companies had over 10 years of exclusive market and could recoup the huge expenses involved in discovering and developing innovative drugs. Recently, generic companies have lost their patience because of the potential for enormous profit that blockbuster drugs promise. They are changing the rules. Now they often legally challenge the validity of patents for a lucrative brand-named drug years before patents expire. Because the stakes are so high, legal fees are dwarfed by potential profits. If the challengers are successful, they are given six months of exclusivity for the generic version of the drug, after which all generic companies are allowed to produce and market the copycat version of the drug, at which point the price inevitably plummets.

A patent is a contract between an inventor and the government whereby the inventor discloses the discovery so others can learn from it. Meanwhile, the government gives the inventor the exclusive right to prevent others from making, using, or selling their invention for a certain period of time. Prescription drugs are protected by patents that give their owners the right to exclude others to make and market the drug for up to 20 years, although they usually must spend at least part of that time gaining federal approval. This patent protection enables the drug maker that discovered the drug to earn back its development costs and make a profit. Otherwise, competitors could immediately make and sell identical versions of the medicine, undercutting the company that invented it. If such a scenario went unchecked, it could lead to the collapse of the pharmaceutical industry, and new drugs would cease to be invented.

In the case of Plavix, the stake was astronomical. Plavix was the world's second best-selling drug, with $6.2 billion in sales in 2005. The first generic company to attack Sanofi's patent on Plavix was the largest Canadian drug company, Apotex in Toronto.

Apotex is a private company founded by Bernard "Barry" C. Sherman. Sherman graduated from the University of Toronto and obtained his Ph.D. in rocket science (I'm not joking) from MIT. As the founder of Apotex, Sherman holds the positions of both CEO and chairman of the board.[42] In essence, Sherman *is* Apotex. Good timing and business savvy made him Canada's 10th richest man with a personal fortune of about $3.7 billion. Sherman is also the biggest litigator in Canada—engaging in 100 court cases simultaneously. As a matter of fact, his litigation expenditure is a full 50 percent of what he invests in research.

In the 1990s, Apotex was at the center of one of the biggest controversies in Canadian science. In 1993, Apotex sponsored a clinical trial run by Dr. Nancy Olivieri at the Hospital for Sick Children in Toronto. The trial was to test a drug called deferi-prone as an iron chelating agent to treat thalassemia, a congenital blood disorder that causes hemoglobin malfunctions. As a conse-quence, iron accumulates in the body to a dangerously high level, especially in the liver and the heart. When Olivieri discovered (objectively or subjectively, it is still of great debate[43]) serious liver scarring in some patients on deferiprone in 1996, Apotex termi-nated her trial and threatened to sue if she ever published the results. Olivieri blew the whistle, and her David-versus-Goliath fight against Apotex attracted international attention. Although Olivieri's results were not later reproduced by other research-ers, and deferiprone was eventually licensed in 47 countries (although not in the United States and Canada), Apotex's rough handling of the Olivieri affair did not win them many friends in the scientific field. Author John le Carré wrote a novel featuring an Olivieri-like heroine: *The Constant Gardener*, which was later adapted as a major motion picture starring Ralph Fiennes and Rachel Weisz.

In 1999, Apotex CEO Sherman was in hot water. The president of the University of Toronto, Robert Prichard, was discovered to be lobbying the prime minister to block legislation that was unfavorable to the generic drug industry. When confronted, Prichard admitted that Sherman told him that if the legislation was passed, he would not be able to fulfill his promise of donations of millions to the university. Prichard later apologized publicly.

Just before Christmas, when he was interviewed by a reporter from *60 Minutes*, Sherman said of Olivieri, "She's nuts."[43] Of course, Olivieri sued for libel.

Patent Litigation

To attack the validity of Plavix's patents, Apotex hired the most aggressive litigation lawyers and began their multipronged attacks in Europe, Canada, and the United States. Of course the stakes are always the highest in the United States because Plavix garners almost half of its sales there. In November 2001, Apotex filed an abbreviated new drug application (ANDA) with the FDA and challenged the validity of patent US4,847,265 issued in 1989.[44,45] They claimed that this patent, which covers the right-handed isomer (i.e., clopidogrel bisulfate), was already covered by Sanofi's 1985 patent US4,529,596, which covers PCR 4099, the racemate.

Let's look at history to get a perspective on the patentability of enantiomers. Although chirality has been known about for more than 165 years since Pasteur's discovery in 1848, chiral drugs did not become popular until the 1980s. True enough, drugs isolated from nature, such as morphine, quinine, and Taxol, are chiral compounds, so there is no chiral separation issue for those drugs. During the second half of the last century, the golden age of the pharmaceutical industry, thousands of synthetic drugs were prepared, tested, and marketed. Among those, many drugs were prepared as racemates, such as fluoxetine (Prozac), gatifloxacin (Tequin), amphetamine (Adderall), sotalol

(Betapace), and albuterol (Ventolin). Separacor, a company near Boston, was overtly founded to take advantage of the chirality of drugs. They took well-known racemate drugs, separated their enantiomers, and tested them. For instance, albuterol was a short-acting β2-adrenergic receptor agonist discovered by GSK. Separacor successfully won the FDA's approval to market the left-handed isomer and sold levalbuterol under the trade name Xopenex. Although never explicitly stated in their commercials, Separacor implied that Xopenex produced fewer side effects than the original racemate.

These days, pharmaceutical companies have most often attempted to evaluate enantiomers of a racemate right after the initial discovery of the racemate, since it is now well known that enantiomers differ in pharmaceutical properties such as potency, efficacy, PK profile, and toxicity.

After Apotex filed an ANDA and began to challenge the validity of the Plavix patent, Sanofi sued Apotex for infringement of the composition-of-matter patent US4,847,265 for the enantiomer clopidogrel bisulfate. But the uncertainty of the litigation had a negative impact on investors' confidence. As a result, Sanofi/BMS chose to settle out of court later on, and an accord was reached between Sanofi/BMS and Apotex in March 2006. There was nothing wrong about this. Settlements between brand-name drug companies and generic drug companies are routinely reached and approved by governmental regulators, including state attorneys general and the Federal Trade Commission (FTC) in the U.S. However, in this case, the March settlement was rejected by the FTC.

Dr. Andrew G. Bodnar was then BMS's senior vice president for strategy and medical external affairs. BMS CEO Peter Dolan dispatched him to carry out the patent settlement with Apotex on behalf of Sanofi and BMS. A new settlement on the Plavix patent was put together in June and submitted to the FTC and state attorneys general again for approval. If approved, Apotex would gain some $40 million. In turn, Apotex agreed not to launch a generic drug until 2011—eight months before the patent protecting the drug was due to expire anyway. One unusual concession was that if the authorities did not approve the deal and Apotex released generic clopidogrel regardless, it would not have to pay triple damages to Sanofi/BMS. The FTC again rejected the settlement at the beginning of August, and Apotex's attorneys informed the FTC that the companies had entered into several side agreements that they had kept secret from the authorities. They alleged that Bodnar made secret assurances to Apotex that Bristol-Myers would not issue its own generic version of the drug to compete with Apotex. What happened next would shake up Wall Street.[46]

As soon as the FTC rejected the August deal, Apotex began to ship generic clopidogrel to the market on August 7, 2006. Despite the government's rejection of the deal, some terms of the agreement remained in effect. Under the terms, Sanofi/BMS had to wait five business days before seeking a federal injunction against Apotex's shipments. *The New York Times* proclaimed "Marketers of Plavix Outfoxed on a Deal."[47] Sherman said that he had never expected the U.S. government to approve the deal but that he had conducted the negotiations in a way that would allow him to push his drug onto the market. Apotex immediately flooded the market with about a six-month supply of clopidogrel in three weeks. A federal judge ordered a halt to Apotex's shipments, but the pills already in the distribution pipeline were not recalled. As a consequence, BMS lost over a billion dollars of sales for Plavix, which explains the unique dip in sales of Plavix in 2006.

As part of a federal investigation, the FBI searched the offices of BMS CEO Peter R. Dolan and Dr. Andrew Bodnar, who was the person who visited Sherman's Toronto office twice to personally negotiate part of the deal. In September, BMS ousted Dolan and its general counsel, Richard K. Willard, for mishandling an effort to negotiate an illegal deal with Apotex.

Finally, the Plavix patent trial was under way on January 22, 2007, in Federal District Court in Manhattan before Judge Sidney H. Stein, who heard the case without a jury. The language that the Apotex lawyers presented was very colorful.[48] They claimed that "these elements were not inventive in the least," because US4,847,265 was anticipated by Sanofi's previous patent US4,529,596. They accused Sanofi of "inventing patents rather than inventing drugs" and said that "Sanofi also comes to this court with unclean hands." They called "Sanofi's delay tactics plainly subversive of the Hatch-Waxman regime." Furthermore, Apotex asked the judge to invalidate US4,847,265 and reject Sanofi/BMS's injunction because of their laches and unclean hands, and so on.

In the end, the Sanofi/BMS lawyers successfully convinced Judge Stein of the validity of US4,847,265. Judge Stein ruled in his "Opinion and Order" that Apotex failed to prove by clear and convincing evidence that the patent was invalid or unenforceable on any of the grounds asserted.[36] Indeed, without separating the two enantiomers, there was no way of knowing which specific enantiomer might contribute to the platelet aggregation inhibiting activity, or the toxicity. Therefore, the therapeutic activity of the right-handed enantiomer clopidogrel was unexpected, as was its tolerance. The judge also rejected Apotex's other frivolous and sensational claims. The conclusion: "Apotex has concededly infringed Sanofi's 265 patent." Judge Stein informed the appellants from both sides that "damages will be set in an amount to be determined through future proceedings."

On October 18, 2011, the U.S. Court of Appeals for the Federal Circuit in Washington ruled that Apotex was required to pay Sanofi and BMS $442 million for infringing the 265 patent.

On the other hand, the court declined to make Apotex pay an additional $107.9 million in interest.

Two Years of Hard Writing

In April 2008, BMS's Bodnar was indicted on federal charges of negotiating the secret pact and then hiding the deal from federal regulators. In June 2009, Judge Ricardo M. Urbina of the U.S. District Court for the District of Columbia sentenced Bodnar to two years of probation during which he was to write a book about his experience connected to the case. Bodnar had to also pay a $5,000 fine. "I would like to see you write a book" so other people "don't find themselves in a similar situation," the judge told Bodnar. "Who knows, it may even be inspirational."[49]

§5.5. Beyond Plavix

Effient and Brilinta

If Ticlid is considered the first generation of thienopyridine blood thinners, and Plavix the second, then Eli Lilly's prasugrel would be the third generation of thienopyridine blood thinners. Prasugrel (trade name Effient) is a me-too drug with the same core structure as clopidogrel. Daiichi Sankyo, Japan's largest pharmaceutical company, came up with prasugrel by modifying clopidogrel's structure. The key difference is that prasugrel has a ketone group to replace clopidogrel's methyl ester (figure 5.9). As a consequence, in vivo, prasugrel does not have the acid metabolite, which is inactive in antiplatelet aggregation.

The modification resulted in a dichotomy. On the one hand, Sankyo was not only able to obtain its intellectual properties through patents, but the resulting prasugrel is more potent

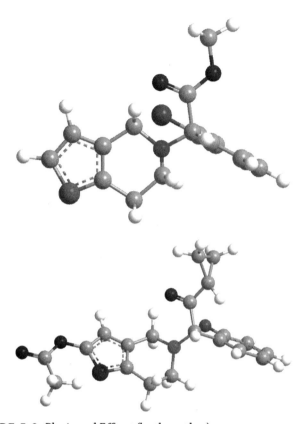

FIGURE 5.9 Plavix and Effient (by the author).

and fast-acting than clopidogrel. Prasugrel reaches its maximum anticoagulant effect more quickly than clopidogrel does. However, prasugrel's inherent higher potency dictates its efficacy and safety endpoints, which include higher incidents of bleeding than clopidogrel.

In the United States, Sankyo chose Eli Lilly as its partner for prasugrel. Lilly ran into some problems during the Phase III clinical trials when they had to stop giving prasugrel to some patients because the dosage needed to be changed. In the prasugrel group, there was a higher incidence of bleeding among people who had previously had strokes, people aged 75 or older, and people who weighed less than 132 pounds. After completion of the trials, prasugrel trod a tortured, delayed path through the FDA. One critic, Sanjay Kaul of Cedar-Sinai Medical Center, was originally named as a voting member of the panel. He was excluded at the last minute after Lilly complained to the agency about his "pre-conceived bias." The agency delayed its decision in June 2008 and again in September. By February 2009, an advisory panel recommended that the FDA approve prasugrel. The drug was finally approved in July 2009 after 18 months of delays.[50] Prasugrel (Effient) was approved for the reduction of thrombotic cardiovascular events (including stent thrombosis) in patients with acute coronary syndromes (ACS) who are managed with an artery-opening procedure known as percutaneous coronary intervention (PCI). But Effient has to carry a "black box" warning about its potential for "significant, sometimes fatal, bleeding." It tells doctors not to prescribe the drug for patients with "pathological bleeding" or for those with a history of strokes or mini-strokes. It also says that Effient is "not recommended" for patients older than 75 years of age unless they are especially at risk of heart disease, and it states that underweight people and those on certain medicines are at increased risk for serious bleeding. Due to the bleeding side effects, Wall Street expected that Effient would be a "niche drug," capturing about 15 percent of Plavix's market. However, with Plavix having lost patent protection in 2011 in the United States and in 2013 in Europe, Effient may be able to grab a bigger share of the oral anticoagulant market.

Because Plavix and Effient, both thienopyridines, are ADP-dependent platelet aggregation inhibitors, efforts have been ongoing to find inhibitors that are not thienopyridines. AstraZeneca's ticagrelor (trade name Brilinta) is the fruit of such endeavors. Like Plavix and Effient, Brilinta blocks ADP receptors of subtype P2Y12. Unlike Plavix and Effient, Brilinta's platelet aggregation inhibition activity is reversible. Moreover, unlike Plavix, it does not need hepatic activation, which could reduce the risk of drug-drug interactions. In May 2009, AstraZeneca reported a large study, including more than 18,000 patients, that compared how Brilinta and Plavix reduced the risk of a composite of heart attacks, strokes, and death from cardiovascular causes.[51] Brilinta seemed to be more effective at preventing heart attacks and strokes than Plavix, although it will be a while before definitive results become known.

In July 2011, the FDA approved ticagrelor (Brilinta) for marketing in the United States. Interestingly, since the Plavix patent expired in May 2012, the field of blood thinners has undergone a dynamic influx and redistribution of sales.

Portola Pharmaceuticals is a privately held biotechnology company based in South San Francisco. Portola's elinogrel is also an ADP-dependent platelet aggregation inhibitor whose core structure is quinazoline-sulfonyl-urea, completely unrelated to thienopyridines. Elinogrel might be faster than Plavix, and its effects might be more reversible. Novartis purchased the right to elinogrel for $75 million in 2009 when it was still in Phase II clinical trials.[52] Novartis also pays for elinogrel's Phase III clinical trials. Portola will be eligible for up to $500 million in milestone payments from Novartis and a double-digit royalty on sales of the drug.

In July 2009, Merck & Co. obtained worldwide rights to codevelop and commercialize an experimental drug in a crowded race to replace the anticoagulant warfarin.[53] Merck paid closely held Portola, Inc., $50 million to license the factor Xa inhibitor betrixaban. It is in Phase II human studies to prevent the formation of blood clots in a variety of conditions. Portola may receive additional payments totaling as much as $420 million if the drug meets certain milestones. Portola did very well in the anticoagulant field with its deal with Novartis on elinogrel and the deal with Merck on betrixaban.

Xarelto and Eliquis—Factor Xa Inhibitors

Heparin, discovered in 1916, was the first anticoagulant. By the end of the last century, we learned enough to understand that most parts of heparin are just decorative, as far as anticlotting activities are concerned. This knowledge led the company Jean Choay (absorbed by Sanofi in 1983) in France to synthesize pentasaccharides with only five key sugar units that are required for its activities. Among those, the first commercial sulfated pentasaccharide fondaparinux has been on the market since 2001. With a half-life of 17 hours, fondaparinux has to be given once daily by subcutaneous administration. The other pentasaccharide idraparinux, the hypermethylated fondaparinux, has a half-life of 80 hours, the longest-acting pentasaccharide. Idraparinux needs to be given only once a week. The Van Gogh studies[54] solidly demonstrated its utility in treating both acute VTE (venous thromboembolism) and extended thromboprophylaxis.

Both fondaparinux and idraparinux are selective indirect factor Xa inhibitors, which act through reversible high-affinity binding to antithrombin, with very little effect on platelet aggregation. Factor Xa is a trypsin-like serine protease that plays an essential role in the coagulation cascade. Hirudin and bivalirudin have been found to be factor Xa inhibitors as well, although they have to be administered subcutaneously. Oral factor Xa inhibitors would be better anticoagulants, and the pharmaceutical industry has been hard at work to find them for the last decade. At one time or another, almost every drug firm has had a factor Xa inhibitor program.

Initially, almost all the factor Xa inhibitors were dogged by the necessity of the basic amidine group, the same group also in AstraZeneca's melagatran. In order to achieve good oral bioavailability, it was critical to find replacements less basic than amidine that still retained potency. One of the frontrunners in this very competitive field was Bayer with their rivaroxaban, discovered by chemists in Wuppertal, Germany. At the outset of the project, Bayer medicinal chemist Susanne Roehrig and her colleagues started as everyone else did with an amidine compound.[55] Just like everyone else, they made an effort to steer away from amidine. One thing led to another, and they replaced amidine with morpholine and chose their initial core structure with oxazolidinone. As the project progressed, slowly but surely, the Bayer factor Xa inhibitors looked more and more like Pharmacia's antibacterial drug linezolid (trade name Zyvox). As a matter of fact, Bayer's lead compound, BAY 59-7939, contains all the fragments of Zyvox except two sites of modification at the two tails of the molecule. Nonetheless, the drug is an orally active, direct factor Xa inhibitor that has a high selectivity and good pharmacokinetics in rats and dogs. Encouraged, Bayer moved the compound into clinical trials and gave it the generic name rivaroxaban. In July 2008, the European Medicines Agency approved licensing of rivaroxaban for use to reduce the risk of blood clots in the weeks following knee- and hip-replacement surgeries. Bayer sells it in Europe with the trade name Xarelto.

The drug had a tougher time with the FDA. Bayer chose Johnson & Johnson as their U.S. partner for Xarelto.[56] Despite a 15-2 vote by an advisory panel in favor of the experimental anticlotting drug rivaroxaban in March 2009, the FDA issued Johnson & Johnson a "complete response letter" (similar to a nonapproval letter) in May. Lingering concerns about a risk of harm to the liver likely prompted the FDA to take more time to review data from a recently completed study called ATLAS. The agency also asked for European market surveillance data. Eventually, the FDA approved Xarelto in July 2011 for the prevention of venous thromboembolism in adult patients undergoing elective hip- and knee-replacement surgeries.

The second frontrunner in the field of factor Xa inhibitors is Bristol-Myers Squibb's apixaban (Eliquis). Medicinal chemists at DuPont Merck initially invented the molecule razaxaban, which became a BMS compound when it acquired DuPont Merck in 2001 for $7.8 billion. For razaxaban, the replacement of amidine was achieved using a primary amide. It was a highly potent reversible anticoagulant with a high affinity and high degree of selectivity for factor Xa. However, the presence of an amide group made the drug liable to metabolism in vivo. BMS improved the bioavailability by eliminating the amide and arrived at apixaban.[57] By March 2005, BMS decided to discontinue razaxaban to concentrate on the development of apixaban. After Phase I and II studies, BMS sought Pfizer to be their codevelopment partner in April 2007. Pfizer paid $250 million upfront, 60 percent of the development cost, and up to $750 million in milestones. Development, expenses, and profits were to be shared equally.

In May 2011, the European Commission approved apixaban (Eliquis) in the 27 countries of the European Union for the prevention of venous thromboembolic events in adult patients who have undergone elective hip- or knee-replacement surgeries. On the last working day of 2012, the FDA also approved apixaban (Eliquis) for the treatment of arterial fibrillation.

Conquest of Pain

Analgesics: From Morphine to Lyrica

The easiest pain to bear is someone else's.

ANONYMOUS

To live is to endure pain has been understood by almost everybody who is mature enough to gain some philosophical perspective on life. *C'est la vie!* as the French would say. Indeed, pain existed before the dawn of humanity—some research suggests that even plants respond to pain.

According to ancient Greek myth, Prometheus stole fire from Olympus to give it to mortals. Zeus punished him by chaining him to a rock and having a great eagle feast on his liver daily, inflicting unbearable agony. Zeus also sent Pandora to Earth, unleashing pain (one of the items in Pandora's box) and many evils as a vengeance to mankind. Without an understanding of pain, our ancestors resorted to many measures to ease pain; some were successful to some extent, and some were completely futile. Witches and shamans were sought out to exorcise pain from the body. From a psychological perspective, they might be effective for some believers. The hypnotizing technique reached its crescendo in the 18th century in France when Monsieur Anton Mesmer "*mesmer*ized" many French citizens, liberating them from their pains.

As civilization progressed, alcohol became more and more a universal painkiller after it was observed that drunkards were oblivious to pain. Chinese surgeon Hua Tuo (115–205 AD) gave his patients an effervescent powder (possibly cannabis) in wine that produced numbness and insensibility before surgical operations.

Another ancient invention in Chinese medicine was the use of acupuncture to ease pain. Acupuncture, now an increasingly popular treatment for persistent as well as intermittent pain, is thought to work by increasing the release of endorphins, chemicals that block pain signals from reaching the brain. A recent survey by the National Institute of Health (NIH) indicated that acupuncture showed efficacy in adult postoperative pain, chemotherapy nausea and vomiting, and postoperative dental pain.[1] There is no doubt that acupuncture works for some patients' minor pain, through either physiological or psychological means, or both. During the hype of the Great Culture Revolution (1966–1976), it was even claimed that major operations were carried out using acupuncture without any other anesthetics. The commemorative postage stamp from 1975 (figure 6.1) depicts

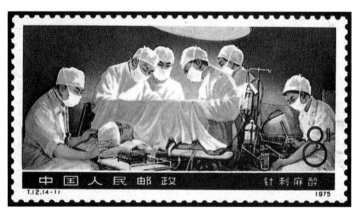

FIGURE 6.1 Acupuncture as anesthesia for surgery. © China Post.

open-heart surgery using only acupuncture for anesthesia. Fact or fiction? You be the judge.

§6.1. Aspirin and Other NSAIDs

Today, pain is the number-one reason for Americans to visit a physician. Analgesics, also known as palliatives, are among the most popular medicines.

Nearly every adult has taken non-steroidal anti-inflammatory drugs (NSAIDs). Aspirin, discovered by Bayer's Felix Hoffmann in 1897, was the first NSAID.

Aspirin's genesis traces back to the bark of willow trees, which were used medicinally as early as 1500 BC. The *Ebers Papyrus* referenced the willow's medicinal properties in general, and its treatment of rheumatism in pregnant women in particular. In Greece, the father of medicine, Hippocrates, recommended using the bark of the willow tree as an analgesic in his writings. In England, Reverend Edward Stone in 1753 experimented with the extraordinarily bitter willow bark in treating ague (fever from malaria) and other disorders with satisfactory results. Five years later, he communicated his experience to the president of the Royal Society.

About two hundred years ago, the bitter tasting salicin, the precursor of salicylic acid, was isolated as yellow crystals from many plant sources including willow (*salix alba vulgaris*), wintergreen, poplar, and bark. When salicin is ingested, it is hydrolyzed into glucose and salicylic alcohol, which is then oxidized into salicylic acid in the stomach. Salicylic acid is the active principal ingredient in aspirin.

On October 10, 1897, Bayer's chemist, Felix Hoffmann, prepared aspirin in a purer form using an improved process. Hoffmann gave some of his aspirin to his father, and it worked well in relieving arthritis pain without severe gastrointestinal (GI) side effects, unlike sodium salicylate, which had bothered his father's stomach. Further testing also confirmed aspirin's superior attributes. However, his manager Heinrich Dreser and the head of Bayer, Carl Duisberg, were more enamored with heroin, another Hoffmann invention. They were against aspirin's development and marketing because they both believed all salicylic acids had detrimental effects on the heart (now we know that aspirin does have beneficial effects for the heart). Exasperated by the stonewalling, Hoffmann's colleague Arthur Eichengrün tested aspirin on himself and did not suffer any heart maladies. He smuggled some samples of aspirin to doctors, and the response was so positive that the company management had no choice but to market it. And as the saying goes, the rest is history.

Interestingly, although aspirin was the most commercially successful drug in the *Guinness Book of Records* in 1950, its mechanism of action (MOA) was not known until 1971 when John R. Vane discovered that aspirin works by preventing the synthesis of prostaglandins by blocking the cyclooxygenase enzyme. In 1982, the Nobel Prize in Physiology or Medicine was awarded to Vane, Bergström, and Samuelsson for their discoveries concerning prostaglandins and related biologically active substances (see chapter 5).

Knowing aspirin's MOA, it is now easier to appreciate how aspirin works. Pain is caused when several endogenous substances are injected into tissues. Those compounds include histamine, serotonin, proton, bradykinin, and prostaglandins E_2 and I_2. Aspirin can relieve pain by blocking the function of cyclooxygenase, thus blocking the production of prostaglandins E_2 and I_2. In terms of medicinal chemistry, aspirin may be defined as a suicide inhibitor of the cyclooxygenase enzyme because it actually forms a covalent bond with the enzyme and thus inhibits its function and loses its own activity at the same time.

According to the contemporary definition, NSAIDs are drugs that exert their anti-inflammatory effects via blockage of prostaglandin synthesis. According to this definition, diflunisal (Dolobid), indomethacin (Indocin), ibuprofen (e.g., Advil), and naproxen (Aleve) are NSAIDs, but acetaminophen (e.g., Tylenol) is not. Acetaminophen is biologically active as an analgesic possibly through inhibition of cyclooxygenase-3 (COX-3), a subtype of the cyclooxygenase enzyme not identified until just a few years ago.

§6.2. Celebrex and COX-2 Inhibitors

Celebrex and other COX-2 inhibitors are the latest NSAIDs. By the mid-1980s, evidence began to emerge that there are two subtypes of cyclooxygenase. Slowly but surely, the two isoforms of cyclooxygenase were deciphered and isolated thanks to the elegant work of several groups of scientists including Philip Needleman at Washington University in St. Louis (he later moved to Monsanto to capitalize on his discovery); Daniel L. Simmons at Brigham Young University; Harvey R. Herschman at UCLA; and Donald Young at the University of Rochester. In 1991, the novel isoform of the inducible cyclooxygenase in the arachidonic acid/prostaglandin pathway was officially named cyclooxygenase II (COX-2).

The two subtypes of cyclooxygenases, COX-1 and COX-2, function differently. COX-1 is mainly responsible for normal physiological processes such as protecting the gastric mucosa and maintaining dilation of blood vessels. COX-2 is localized mainly in inflammatory cells and tissues and becomes active during acute inflammatory response. Therefore, in general terms, COX-1 is the good enzyme, and COX-2 is the bad one. Consequently, selectively blocking COX-2 inhibitor would provide an anti-inflammatory drug by selectively inhibiting prostaglandin production. Meanwhile, leaving the good enzyme COX-1 alone would reduce adverse gastrointestinal and hematologic side effects.

Applying this novel concept, Monsanto came up with the first COX-2 selective inhibitor celecoxib under the leadership of Philip Needleman. The medicinal chemistry team was led by John A. Talley in St. Louis. Using an HTS hit as the lead and through intensive structure-activity relationship (SAR) investigations, rofecoxib was first synthesized on October 4, 1993, at G. D. Searle, a division of Monsanto. Pharmacia, which acquired Monsanto's pharmaceutical division in 1998, began co-marketing celecoxib with Pfizer in June 1999 with the brand name of Celebrex. The partnership sowed the seeds for Pfizer's acquisition of Pharmacia in 2003. Indeed, mergers and acquisitions (M&As) in pharma were happening at such a dizzy pace, it was hard for anybody to keep track of which company is which and which company "swallowed" which other company.

At the end of 1999, Merck received approval from the FDA to market their version of a COX-2 selective inhibitor, rofecoxib (Vioxx). Vioxx was discovered by a team led by P. Prasit at Merck-Frosst in Montreal, Canada. Merck acquired Charles E. Frosst in 1965, and Merck-Frosst was also the birthplace of the blockbuster drug montelukast (Singulair) for the treatment of asthma and chronic obstructive pulmonary disease (COPD).

Both Celebrex and Vioxx quickly became blockbuster drugs for treatment of pain associated with both osteoarthritis (OA) and rheumatoid arthritis (RA). In 2003, Bextra, a follow-up of Celebrex, was FDA-approved and co-marketed by Pfizer and Pharmacia, the name of the newly formed company from a merger between Pharmacia, Upjohn, and Monsanto (the parent company of Searle). The financial windfall that Celebrex and Bextra promised to bring cemented the newly formed Pharmacia's fate. Pfizer gobbled it up in 2003 to secure the exclusive right to the COX-2 franchise, just like it did three years earlier with Warner-Lambert, its Lipitor marketing partner, in 2000.

Sadly, in Merck's VIGOR clinical trials published in 2004, there was a fivefold increase in myocardial infarction among patients taking Vioxx compared to patients taking naproxen. In September 2004, clinical trials showed that Vioxx increased the risk of myocardial infarction and stroke. Merck voluntarily withdrew Vioxx from the market on September 30. The stock market reacted violently, wiping out $30 billion of the value of Merck stocks in a matter of a few months. In 2009, the storied Merck-Frosst site in Montreal was regrettably closed, along with many drug firms in the Montreal area, an unfortunate outcome partially attributed to the demise of Vioxx.

In April 2005, acting against the vote of its advisory panel, the FDA asked Pfizer to "voluntarily" withdraw Bextra. Presumably, that was when the political wind blowing toward the conservative extreme forced the FDA to err on the cautious side.

Today, Celebrex is the only COX-2 selective inhibitor remaining on the market.

§6.3. Morphine and Opiates

By far the most important class of drugs for treating moderate to severe chronic pain is the opiates, including morphine and morphine-like drugs. Especially for chronic, continuous pain, a slowly released opiate like oxycodone (Oxycontin) is preferred. These drugs minimize or eliminate the hills and valleys of pain and reduce the medication that patients need. A fast-acting one like Percocet (oxycodone with acetaminophen) is also available to treat pain.[2]

Narcotics such as opium, laudanum (coined by Paracelsus in 1524 from the Latin root *laudare*, meaning "to praise"), and henbane have been used for millennia to relieve many ills. Opium is extracted from the bulbs of *Papaver somniferum* (figure 6.2), which was cultivated in the Mediterranean area as early as 5000 BC. Opium's medicinal properties were evidently discovered before the beginning of recorded history.[3] Because making opium did not require fermentation, it is likely that the use of opium predates alcohol, making it one of the first medicines of humans. George Washington and Florence Nightingale are among numerous historic figures who used laudanum regularly. Morphine, the principal component of opium responsible for its analgesic and euphoriant properties, was isolated by then 20-year-old German pharmacist Friedrich Adam Sertürner in Hannover in 1803. He named the white crystalline substance morphine after the Greek god of dreams Morpheus. Morphine's profound analgesic effect makes it one of the most useful painkillers even today.

FIGURE 6.2 *Papaver somniferum*, from which opium is extracted. Photo by author.

Sertürner was also the first person to prove that morphine was basic, which made it an alkaloid—as in alkaline. Before that, it was thought that all organic compounds were acidic. In 1973, Solomon H. Snyder and his student Candace Pert of Johns Hopkins University reported the characterization of the pharmacological target for morphine, the μ-opiate receptor. Although morphine binds to all the opiate receptor subtypes, including the δ and κ receptors, its analgesic activity is mostly mediated by the μ-opiate receptor.

Interestingly, the human brain naturally contains endogenous morphine-like analgesics. The endogenous opioids and enkephalins (both methionine-enkephalin and leucine-enkephalin) are two neuropeptides consisting of five amino acids each. They differ only in one terminal amino acid: methionine versus leucine. It turns out that evolution has endowed us with our own painkillers. This may explain why pain treatment has one of the highest rates of placebo effect: 30 to 50 percent. Pharmacologically, the placebo effect is probably generated by the endogenous opioids.

British chemist Alder Wright first synthesized diacetylmorphine in 1874. While Wright was experimenting with opium derivatives at St. Mary's Hospital in London, he obtained a white crystalline solid from the reaction between morphine and acetic anhydride. Two decades later, a chemist at Bayer AG in Germany, Felix Hoffmann, repeated Wright's experiment and transformed morphine to diacetylmorphine in 1897 using a similar process to the one he used in converting salicylic acid to aspirin. The head of Bayer's pharmacology department, Heinrich Dreser, first tested the drug on frogs and rabbits. He then tried it on himself and volunteers from the adjacent Bayer Dye Factory. The workers felt that the drug made them feel so "heroic" that Dreser named it heroin. Bayer promoted heroin as a safe substitute for morphine, addiction to which was becoming a growing plague in Europe at the time. From then on, heroin received widespread acceptance and was sold over the counter for all imaginable maladies across Germany, followed by Europe and America. Needless to say, heroin abuse (figure 6.3) began since the first days it was on the market.

Other alkaloids found in opium with analgesic activity mediated by the μ-opiate receptor include codeine and thebaine. Codeine, a minor constituent of opium, was isolated in 1832. It is methylated morphine. Therefore, it may be considered a prodrug of morphine. When ingested, codeine is demethylated to morphine in vivo. Today the majority of legitimate morphine production is used to convert it to codeine, because codeine is approximately tenfold less powerful as morphine and not likely to cause addiction. (In the author's post-periodontal-operation experience of codeine to ward off pain, no addiction ensued.)

Thebaine is named after the Egyptian city of Thebes, which was so famous for its poppy fields in 15th-century BC that Egyptian opium was known as Thebic opium. Thebaine, bis-methylated morphine, is also the source of a number of semisynthetic opiate drugs also used as analgesics, such as oxycodone (Oxycontin), oxymorphone (Opana), and etorphine (Immobilon). The latter

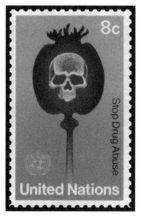

FIGURE 6.3 Stop Drug Abuse. © United Nations.

is at least 100 times more potent than morphine and is fatal to humans, so it is only used in tranquilizing large animals like elephants.[4]

In addition to μ-opiate agonists, the μ-opiate antagonists, naloxone (Narcan), naltrexone (Revia), and nalmefene (Revex), have found medical utility as well. While naloxone is used to reverse opiate-induced side effects from overdoses such as those that occur in heroin addicts, the latter two drugs are used in controlling alcohol abuse, as they are postulated to lessen the euphoria reward associated with alcohol consumption, encouraging abstinence.

In addition to the naturally occurring opioid drugs, there are several synthetic analgesics that bind to the same receptor target. Meperidine (Demerol) was originally synthesized at I. G. Farben in 1932 and expected to be an antispasmodic in analogy to the muscarinic agents after which it was modeled. But designs sometimes go awry, and in this case, meperidine's potent analgesic activity was discovered later in 1940, although its mechanism of action (MOA) was established much later as being the same as the opioids, with agonist activity at the μ-opiate receptor. Although Demerol's efficacy supposedly came without the side effects of the other opiate drugs, its potent affinity for the dopamine and norepinephrine transporters gives it a pharmacological activity similar to that of cocaine, and it actually carries a greater potential for addiction. In addition, its drug-drug interactions (DDIs) can be serious, and hence it is no longer widely used clinically.

§6.4. Fentanyl and Opioids

Morphine is a complicated natural product, rendering its synthesis highly challenging. A pentacyclic alkaloid, it has five rings and five continuous chiral carbon centers. Its structural elucidation was attempted by many organic chemistry luminaries for decades, including Justus von Liebig, Robert Pschorr, August W. Hofmann, Ludwig Knorr, Julius von Braun, Clemens Schöpf, and Heinrich Wieland, a chemistry "Who's Who" for the later 1800s. The correct structure was not known until it was first proposed by Sir Robert Robinson at Oxford University in 1925 and its total synthesis was achieved by Marshall Gates at Bryn Mawr College in Pennsylvania in 1952. Despite his elegant and well-known achievements, Gates remained unaffected by his success and shunned public attention, preferring to quietly continue his work as a much-respected teacher and researcher initially at Bryn Mawr and later at the University of Rochester.[5]

Eventually, morphine's stereochemistry was confirmed by means of X-ray crystallography by Nobel Laureate Dorothy Hodgkin of Oxford in 1955.

Making morphine analogues is really challenging; but the good news is that medicinal chemists can always do what they do the best: structure-activity relationship (SAR) explorations. Namely, chemists can chop off fragments of morphine, simplify the molecule, and zero in on the pharmacophore—the minimal fragments of the molecule needed to exert its pharmacological effects. That's what medicinal chemists did with morphine.

Before describing the exploits of the medicinal chemists, a terminology clarification for opiates and opioids is in order. Although opiates and opioids are often interchangeable, opiates are the natural alkaloids isolated from opium poppy and the narcotics made from semisynthetic routes using the natural alkaloids as the starting materials. Opioids, on the other hand, are synthetic drugs that mimic the pain-relieving properties of the natural substances. Therefore, we shall refer to the synthetic painkillers described below as opioids.

In the 1940s, Eli Lilly chemists discovered that not all five rings in morphine were needed to be endowed with the painkilling effect. As a matter fact, there is no need for a single ring at all to maintain the potency. The fruit of their labor resulted in an acyclic (no rings) analgesic propoxyphene (Darvon). Although widely used, Darvon was recently removed from clinical practice due to its limited efficacy and history of troublesome side effects. Methadone (dolophine) was discovered by scientists at I. G. Farben as a remedy for Germany's opium shortage during World War II and was marketed in the United States by Lilly. It eventually became a treatment for heroin detoxification, although it has not been a satisfactory solution to the problem. As a powerful analgesic, methadone cancels out the euphoria of heroin and eases withdrawal. The problem is that methadone is also addictive, which really puts addicts in a conundrum.

A breakthrough came in 1960, when fentanyl (Duragesic) was synthesized at Janssen Pharmaceutica in Belgium.

Paul A. J. Janssen (1926–2003) was an extraordinary chemist. In 1948, he entered the University of Ghent to study with Corneille Heymans, a Nobel Laureate in medicine. But the first classes were so easy for him that he left medical school for a while and decided to visit the United States instead. He had enough money to pay for his voyage to New York, but little left over. He solved the problem by playing chess at the Manhattan Chess Club, where he beat everyone else and won four or five games an evening to make the $1.50 a day that he needed to sustain himself.

In 1951, Paul Janssen duly graduated from Ghent. After spending 18 months of mandatory military service as a physician stationed in Germany, he joined his father's company and became the head of research in 1953. He thought nothing of working six days a week from 6 a.m. to midnight, and he expected the same from his employees. Thankfully, when he got married in 1957 and left work at 7 p.m., so could his employees.

Gratifyingly, their hard work paid off. They developed two safe and efficacious drugs every year, a remarkable feat that has rarely been duplicated anywhere else. In terms of chemistry, they found many successes by incorporating a piperidine in their molecules.

Piperidine: The "Enchanted" Ring

A few words are warranted about why the piperidine ring has been such a "ring of success."[6] Today, it belongs to one of the "privileged" structures—structures that have showed up in many drugs. Nitrogen-containing drugs are prevalent in the cornucopia of all drugs. As a matter of fact, the majority of drugs contain at least one nitrogen atom. Why is that?

Evolution played an important role in determining what kinds of molecules our bodies can absorb. There are numerous reasons why nitrogen-containing drugs are so popular. First of all, amines are often involved in a drug's binding interactions with its target. Second, amines have a good balance of the dual requirements of water and fat solubility. Last but not least, amines are partially ionized in the blood, and they easily equilibrate between their ionized and non-ionized forms. As a consequence, they can cross cell membranes in the non-ionized form while the ionized form gives good water solubility and permits strong binding interactions with the target binding site.

Paul Janssen became interested in narcotic analgesics as early as 1953 when his professional research work began. He and his chemists were also heavily influenced by an existing drug, meperidine (Demerol) (figure 6.4), first synthesized in Germany in 1939 during

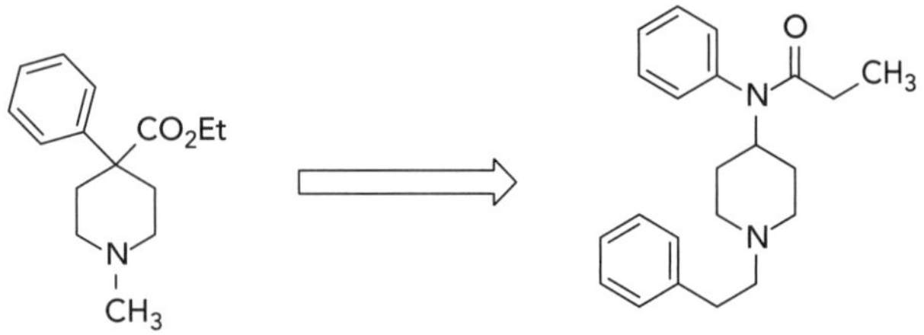

Meperidine (Demerol)   Fentanyl (Duragesic)

FIGURE 6.4 From Demerol to Duragesic.

World War II. The drug is stripped from fragments of the morphine molecule that do not contribute to its analgesic properties.

Their first success came when they discovered R951, a piperidine ring substituted with a propiophenone side chain. This acyclic compound had a high affinity for the μ-opiate receptor based on the 4-substituted piperidine pharmacophore. This laid the foundation for a better understanding of the SARs of narcotic analgesics and stimulated interest in developing compounds with even greater potency and safety margins. R951 became dextromoramide (Palium) and was on the market in 1956. Three additional piperidine-containing analgesics came in succession: phenoperidine in 1957, piritramide in 1960, and benzitramide in 1961.

Their most popular analgesic, fentanyl (Duragesic) (see figure 6.4), was prepared in 1960. It was 100 times more potent than morphine. Even today, fentanyl is widely used as an analgesic and anesthesia. It has been used to treat breakthrough cancer pain. Because some terminally ill patients might have difficulty swallowing tablets, fentanyl is often administered in the form of lollipops, skin patches, or lozenges for acute breakthrough pain. Occasionally, skin patches could be difficult when gauging the dosage. Anecdotally, a man died of overdose when he applied seven fentanyl patches to his body.

Using his favorite 4-substituted piperidine pharmacophore to model its features, Janssen discovered two far more potent analgesics, sufentanil (Sufenta) and alfentanil (Alfenta).

Over the years, Janssen and his colleagues introduced approximately 80 drugs, including fentanyl and risperidone, that proved useful in human, botanical, and veterinary medicine, unprecedented by any other individual. He built the small drug firm founded by his father into a pharmaceutical giant, which merged with Johnson & Johnson in 1961. He published more than 850 papers, held more than 100 patents, and delivered more than 500 lectures all over the world in Dutch, English, French, German, and Spanish—five languages that he had mastered with fluency. Sadly, Janssen died suddenly on November 11, 2003, while attending a conference in Rome.

Paul Janssen (figure 6.5) performed at his peak during 50 years of intense pharmaceutical research activity. From 1953 to 2003, he discovered over 80 pharmaceutical compounds in various fields of pharmacology.

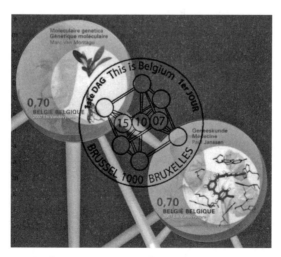

FIGURE 6.5 Paul A. J. Janssen. © Belgium Post.

Regrettably, fentanyls, as a family of very potent narcotic analgesics, have become an increasingly important source of abused substances although they are totally synthetic. Fentanyl itself acquired the street name "China White" among junkies. The most potent analogues, the 3-methyl- and β-hydroxy-fentanyls, may be up to 1,000 times as potent as heroin, but they are not chemically related to the opiates and therefore not detectable by conventional narcotic screening tests.

§6.5. Effexor and Other Antidepressants

Many antidepressants, especially serotonin and norepinephrine dual reuptake inhibitors (SNRIs) like Wyeth's venlafaxine (Effexor) and Lilly's duloxetine (Cymbalta), are now also used as painkillers. At first glance, the link between antidepressants and pain relievers may not so obvious. Here, I offer at least two possible explanations. First, pain is an extremely complex pharmacological process, during which many neurotransmitters are released. Chief among them are serotonin, norepinephrine, dopamine, and acetylcholine. SNRIs block the reuptake of both serotonin and norepinephrine. Therefore, they may help to relive both depression and pain.

The second reason why antidepressants also relieve pain is probably more psychological. We all know that depressed individuals have a lower tolerance level for pain. If the same degree of pain is inflicted on a healthy person and a depressed person, the person with depression may subjectively perceive more pain. A happy and healthy person can tolerate more pain. Therefore, alleviating depression will help to alleviate pain.

Now is a good time to introduce a common, yet quite controversial, syndrome called fibromyalgia syndrome (FMS), for which SNRIs are effective in combination with specific pain-relieving drugs.

Fibromyalgia is chronic, generalized muscle pain. It affects 5 to 20 percent of the population (6 to 10 million people in the United States), and it is controversial because it is ten times more common in women than men. Fibromyalgia is often triggered by a stressful event such as an injury, emotional trauma, or severe flu.

One of the champions of FMS is Dr. Don L. Goldenberg, a physician and professor at Tufts University. Although he was an expert in rheumatology, after his wife Patty developed fibromyalgia in 1977, FMS became a focus for him.

One of the diagnostic methods is to determine tender points on the body. They usually come in pairs for FMS sufferers. Dr. Goldenberg often cautions his fellow physicians to be careful in performing such a diagnosis—because if you press too hard, every point can become a tender point.[7]

In 2009, the FDA approved Forest Labs' Savella (milnacipran), a selective SNRI for the management of FMS. Savella had been marketed for many years in Europe and Asia as an antidepressant, but was only recently investigated as a potential therapy for FMS. The mechanism by which Savella improves the symptoms of FMS is unclear, but it is speculated that abnormal levels of various neurotransmitters may be pivotal in the disease. Therefore, the selective blockade of the reuptake of serotonin and norepinephrine may be the key to Savella's efficacy in fibromyalgia.[8] Thus, Savella joins Lilly's Cymbalta and Pfizer's Lyrica as frontline treatments for the complex symptoms of FMS.

§6.6. Parke-Davis and Dilantin

It may not be so obvious to many, but antiepileptic drugs have been used to treat chronic pain as well.

Epilepsy, afflicting about two million Americans and 40 to 50 million worldwide (1–2 percent of the population), is a very complex disorder. Manifested by falling, tongue-biting, and recurring convulsive seizures (this is why antiepileptics are also known as anticonvulsants), the etiology of epilepsy is believed to be surges in electrical signals inside the brain. Many famous historical figures are known to have had epilepsy, including Aristotle, Julius Caesar, Napoleon Bonaparte, and Vladimir Lenin.

So far, there is no cure for this disorder, and most drugs to either treat the symptoms or act as prophylactics have been discovered by empirical means. The first really effective antiepileptic was phenobarbital. Before that, potassium bromide (a sedative) was only one of the few drugs available as anticonvulsants. But bromide was not very efficacious and caused many side effects.

Adolf von Baeyer prepared the first barbiturate, barbituric acid, in 1864. It is said that he named the compound after his then girlfriend Barbara. Emil Fischer synthesized phenobarbital in 1903, and his friend Joseph von Mering investigated its hypnotic properties in 1911. One year later, a young German physician, Alfred Hauptman, gave phenobarbital to his epilepsy patients as a tranquilizer. Not only did his patients suffer fewer and lighter seizures, some of them were able to be discharged from institutions and went on to resume employment. Even today, phenobarbital is one of the most venerable antiepileptics still in use. Coincidentally, phenobarbital is also used as a minor tranquilizer, as well as a painkiller.

Phenobarbital was successful, but it still possessed sedative side effects. In 1939, Tracy Jackson Putnam and H. Houston Merritt of Boston City Hospital discovered Dilantin (phenytoin, diphenylhydantoin) as an efficacious treatment for epilepsy. Dilantin was the first anticonvulsant that was devoid of sedative side effects, so it was revolutionary for treating epilepsy.

In the mid-1930s, Putnam and Merritt initiated a collaboration on anticonvulsive drugs. Their approach was quite naïve in comparison to today's sophisticated drug discovery methods.

Trying to improve upon phenobarbital, they combed the Eastman Chemical Company's catalog for other commercially available chemicals containing phenyl groups. After that, Putnam wrote to major drug firms requesting phenyl-containing drugs. The first company that they asked was Parke-Davis—the director of research there, Anthony J. Glazko, was a friend of Putnam's. Parke-Davis sent them 19 different compounds analogous to phenobarbital. Phenytoin was among those 19 drugs, sitting on the shelf because it had failed as a sleeping pill.

At the end, Putnam and Merritt systematically screened over 620 drugs using a cat animal model whose convulsion was obtained by electroshock. What stood out the most was Parke-Davis's phenytoin (diphenylhydantoin), which was not only more potent as an anticonvulsant than phenobarbital but also less soporific. It was revolutionary because it was the first drug that separated the antiepileptic effect from the sedative action. As a consequence, phenytoin had fewer side effects than other sedatives for epilepsy. The discovery of Dilantin (figure 6.6) was lauded as the first major successful medical treatment for epilepsy in the 20th century.[9]

Among the 620 compounds that Putnam and Merritt screened, they found a couple of them that were more potent than phenytoin. Unfortunately, they were too toxic to be useful as drugs.

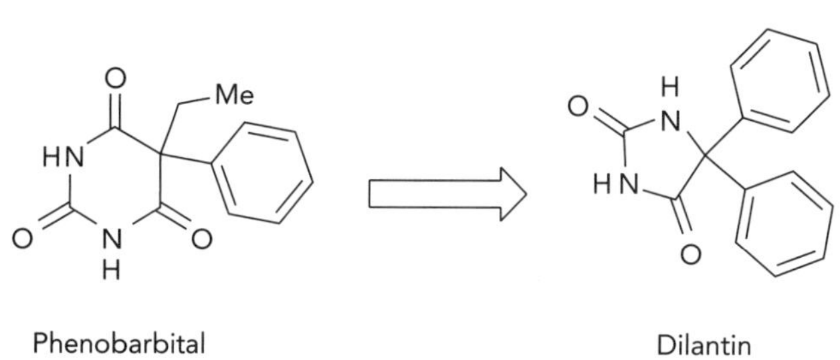

Phenobarbital   Dilantin

FIGURE 6.6 From Phenobarbital to Dilantin.

After Putnam and Merritt's discovery, Parke-Davis began marketing phenytoin for epilepsy with the trade name Dilantin in 1939. Phenytoin was first invented by German chemist Heinrich Biltz in 1908, and Parke-Davis purchased it in 1909. Parke-Davis filed a method-of-use patent, which was granted in 1946 and expired in 1963. After that, Parke-Davis sold it for pennies.

After the emergence of Dilantin, a flurry of anticonvulsants followed, 17 appearing in a time span from 1937 to 1974. In 1954, a chemical-biological research team including L. M. Long, G. M. Chen, and C. A. Miller at Parke-Davis developed Milontin, an anticonvulsant for the control of petit mal epilepsy. In 1957, Parke-Davis introduced another epilepsy drug, Celontin. By then, Parke-Davis's Dilantin, Milontin, and Celontin covered each of the three major types of epileptic seizure. Parke-Davis solidly established its leadership position in the field of epilepsy treatment and gained enormous respect from physicians in the field.

§6.7. Parke-Davis and Neurontin

Building on the success of Dilantin, Parke-Davis came up with an encore in Neurontin. But Neurontin's road to the market was nothing close to a straightforward path.

GABA (γ-aminobutanoic acid), a small amino acid, is a neurotransmitter existing in the human CNS. It was reported at first in the 1950s that GABA existed in the brain. Later on, it was discovered that GABA possibly has inhibitory functions in the CNS. In 1967, it was recognized as an inhibitory neurotransmitter. The discovery of the benzodiazepines Librium (launched in 1960) and Valium by Leo Sternbach at Hoffman LaRoche initiated huge research activities resulting in 50 marketed drugs. It was later determined that benzodiazepines work by modulating the effects of the $GABA_A$ receptor.

GABA regulates neuronal excitability in our brain and muscle tone. Therefore, epilepsy patients may have lower levels of the GABA molecule. However, giving the patient GABA directly is of little help because it is so polar that few molecules would penetrate the blood-brain barrier (BBB) and exert their positive effects. In order to do that, the molecule has to be tweaked to be more greasy (lipophilicity), which was exactly what chemist Gerhard Satzinger thought.

In early 1973, Satzinger was a director of medicinal chemistry at Parke-Davis's German concern in Freiburg, Germany—it was called Gödecke AG before the merger with Parke-Davis after World War II. When he joined Parke-Davis in 1958, he was given the freedom to choose his own area of specialty, as Parke-Davis had a very academic-like culture that was evidently good at fostering innovations. Satzinger put a few novel diuretics on the market during the first decade of his tenure. Later on, he also discovered a couple of analgesics that were launched in Europe but never saw light in the U.S. market.[10] But the crown jewel of his career was a blockbuster drug: Dilantin for the treatment of epilepsy.

Satzinger intended to enhance GABA's lipophilicity by attaching more aliphatic fragments to the GABA molecule. One of the structures that he proposed was to be gabapentin. Basically, he proposed to replace the methylene group with a greasier cyclohexane ring. He drew the structure for gabapentin (initially known as Goe 3450) and a few other similar molecules in July 1973. But it was not prepared by his colleague Dr. Johannes Hartenstein until 1977. A composition-of-matter patent was filed in 1978, and positive results in the epilepsy animal model were obtained from 1979 to 1984.

Unfortunately, Parke-Davis management was not terribly interested in gabapentin as an antiepileptic at first because of strategic concerns. In the late 1980s, Satzinger and his colleagues

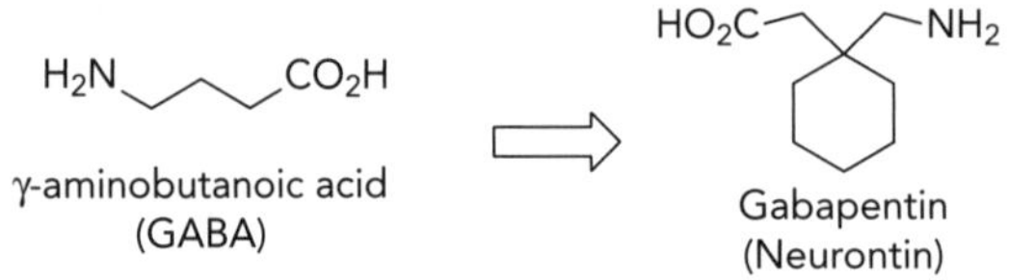

FIGURE 6.7 From GABA to Neurontin.

at Freiburg carried out clandestine development of gabapentin. The clinical trials in epilepsy began in 1984. Parke-Davis management showed real interest in 1985 when positive clinical trials began to surface. The FDA approved gabapentin as a treatment for epilepsy in 1993, and it was finally marketed in 1994 with the trade name Neurontin (figure 6.7).

But there was a problem: The composition-of-matter patent expired during the period of 1993 to 1996. Parke-Davis's patent lawyers were able to obtain a method-of-use (as an adjuvant therapy for treating epilepsy) patent, which was extended to July 2000 thanks to both the Hatch-Waxman act and approval of the drug for pediatric use. In addition, Parke-Davis also patented their manufacturing processes, which also provided additional protection for their Neurontin franchise. In 2001, Neurontin had annual sales of $1.75 billion, a bona fide blockbuster drug. It peaked at $3 billion in 2003.

After 35 years of service, Satzinger retired from Parke-Davis. In an article published in 2001,[11] he was refreshingly blunt in expressing his frustrations and disappointments with management's lack of interest in some of his drugs, including gabapentin. He lamented:

> …One generation later, the medicinal chemist of the early third millennium will note with some nostalgia—at least drug research in which synthetic adventurousness coupled with a bit of pharmacology and physiology and lots of intuition and luck sufficed as ingredients of pharmaceutical discoveries.

Yet he had a sense of humor about the business as well:

> ...I know now that drug discovery requires the same characteristics of a medicinal chemist as of a partner in marriage: patience, endurance, stubbornness and fantasy. After all, there must be a reason why chemists in the United States have the lowest divorce rate of any professional group.

When Satzinger designed gabapentin, he believed that it would be a GABA drug, meaning that it would work by modulating the GABA receptors. Later, it was found to be completely untrue because no evidence could be found that gabapentin interacted with the receptor in the brain.

So how exactly does gabapentin work? Gabapentin's analgesic effects were discovered long after its anticonvulsant effects were known. We explain at the end of this chapter.

§6.8. Silverman and His Theory on GABA Compounds

Richard B. Silverman studied coenzyme B_{12}-dependent enzyme-catalyzed rearrangements at Harvard under the tutelage of Prof. D. Dolphin. After graduating in 1974, he went on to start his academic career at Northwestern University in Evanston, Illinois. One of his research interests was epilepsy.

From 1981 to 1988, Silverman began to design a series of mechanism-based inactivators of GABA aminotransferase as potential anticonvulsants. He rationalized that selectively inhibiting GABA-AT without inhibiting L-glutamic acid decarboxylase would afford a cleaner and better drug. He also appreciated that GABA was such a hydrophilic-charged molecule, he needed to booster its lipophilicity so it could penetrate the BBB.

In 1988, Silverman asked Dr. Ryszard Andruszkiewicz, a visiting scholar from the Technical University of Gdansk in Poland in his group at the time, to prepare a series of 13 simple GABA derivatives, including 3-alkyl-GABAs (figure 6.8) and 3-alkylglutamates.[12] The chemistry was pretty straightforward—the kind that a sophomore undergraduate student would be able to carry out. They treated a series of α,β-unsaturated esters with nitromethane in the presence of tetramethylguanidine as a base (a classical Henry reaction). The nitro products were then converted to the desired 3-alkyl-GABA,[13-15] basically, greasier GABA molecules. In this case, Silverman's approach was not so dissimilar from what Gerhard Satzinger used in designing his gabapentin.

Silverman and Andruszkiewicz tested their 3-alkyl-GABAs using an in vitro assay called GABA aminotransferase, an enzyme that is responsible for catalyzing the catabolism of GABA, the inhibitory neurotransmitter. Indeed, the dozen or so compounds that they made showed desired effects. They published their results in the *Journal of Biological Chemistry* in 1990.

After a patent was filed, the Northwestern University Technology Transfer Program contacted pharmaceutical companies about Silverman's 3-alkyl-GABA compounds. Only two companies replied: Upjohn Pharmaceuticals in Kalamazoo, Michigan, and Parke-Davis Pharmaceuticals in Ann Arbor, Michigan. The two companies were not even two hours apart. While Upjohn

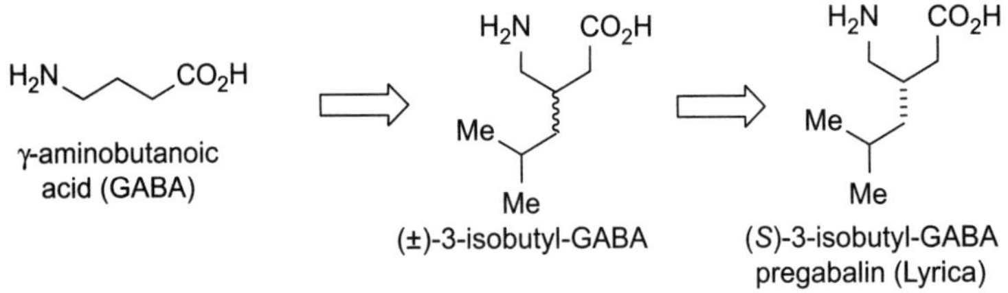

FIGURE 6.8 From GABA to Lyrica.

wanted only the "best" compound, Parke-Davis would test all the analogues. That was logical because of Parke-Davis's long historical association with epilepsy drugs.

Pregabalin was actually the compound that Silverman was least interested in. But after Parke-Davis tested it, it was found to be the most interesting as an anticonvulsant.

§6.9. Parke-Davis and Lyrica

Upjohn tested Silverman's "best" compound, 3-methyl-GABA. They found that (R)-3-methyl-GABA was more potent as a GABA aminotransferase than (S)-3-methyl-GABA. However, when they tested (R)-3-methyl-GABA in the animal model, it was only a weak anticonvulsant. Upjohn quickly lost interest.

With the benefit of history, we know now that Silverman's compounds as anticonvulsants work through a completely different mechanism than inhibiting GABA aminotransferase, as he had originally envisioned.

Meanwhile at Parke-Davis, Charles Taylor and his associate Mark Vartanian tested all of Silverman's compounds, which Silverman himself brought to Ann Arbor. They tested the compounds' potential as anticonvulsants in March 1990 using electroshocks in mice as their initial animal tests. They immediately found that 3-isobutyl-GABA, the compound that Silverman was least interested in, was the most efficacious in that animal model. Then Taylor requested that Silverman retest the compounds in their in vitro GABA aminotransferase assay. At that time, Silverman's group could not reproduce the data they had published in their 1990 paper in the *Journal of Biological Chemistry*.[14]

In 1990, Parke-Davis invited Silverman to give a seminar and discuss his compounds further. The next time Silverman was invited back to Ann Arbor would be more than a decade later, a fact that was a source of bitterness for Silverman.

What Parke-Davis did next with Silverman's compounds was more interesting. It turned out that (±)-3-isobutyl-GABA was efficacious at a very low dose. It almost completely protected the mice from electroshock-induced seizures at a 14.4 mg/kg dose. More interesting, Parke-Davis found that (S)-3-isobutyl-GABA (later known as pregabalin and Lyrica) is more potent than the (R)-isomer. As a matter of fact, it was the most potent anticonvulsant agent they had tested. And the (R)-isomer was not active at all, even at more than 100-fold higher doses.

Meanwhile, medicinal chemists led by Po-Wai Yuen accomplished the first enantioselective synthesis of pregabalin in 1991. Later on, Parke-Davis's Process Chemistry Department in Holland, Michigan, developed an even more robust synthesis for clinical trial use and for manufacturing purposes thereafter.

Because Silverman published his results in 1989–1990 without any patents, it initially made the lives in Parke-Davis's Patent Department quite challenging. It was not until September 2000 when the composition-of matter patent was granted. This turned out to be of great advantage to the drug company because the patent would not expire until September 2017. If the patent had been filed when the Silverman discovery was first made, the patent life would have been 10 years shorter.

Animal pharmacokinetics and metabolism investigations were carried out in 1992, and animal toxicology was studied for the next two years.

An investigational new drug (IND) application was filed at the end of 1995, and pregabalin progressed through clinical trials with flying colors. In 1997, the first Phase I clinical trials were conducted with hospitalized epilepsy patients. Later, it was found that pregabalin had efficacy for both anxiety and pain.

Because pregabalin worked similarly to gabapentin (although the former is a lot more potent), Parke-Davis management had pretty high confidence in the drug. They put a lot of pressure on Michael Pool, the lead clinician for pregabalin, to go all the way with the compound. The large Phase III trials included epilepsy, anxiety, and neuropathic pain, as a matter of fact, the two major types of neuropathic pain in diabetes (peripheral and autonomic). In reality, there were four major clinical trials going on at the same time. Over 2,000 patients participated in these trials. The physicians taking part in the trials had high levels of confidence as well because many of them had experience with Neurontin.

In 2000, Parke-Davis was taken over by Pfizer, which continued pregabalin's clinical trials.

A hiccup in 2001 scared a lot of the researchers working on pregabalin. Although pregabalin worked well without much toxicity for many animals, it reduced breathing in mice on high doses. Two years later, vascular tumors were observed in mice, although rats and all other larger animals did not appear to have this problem. Not surprisingly, the FDA immediately halted the clinical trials on humans. Parke-Davis scientists had to go back to the drawing board, so to speak.

It turned out that in the smaller mice, large doses of pregabalin caused breathing movements, which in turn increased their body acidity as indicated by a decrease of pH values. That was how the vascular tumors were generated in mice. However, this was not an issue with all the other, larger animals in the trials, including humans, so the clinical trials were allowed to continue after the FDA staff was convinced by the solid data generated from the investigations.

In 2003, Pfizer filed the NDA (new drug application) with the FDA, and the drug was approved in the United States for marketing in 2004 with the trade name Lyrica. It turns out that choosing a trade name for a drug is not a walk in the park.[16]

A good drug trade name should be least likely to cause dispensing errors. So, a name that is similar to an existing drug is immediately ruled out. It should also be easily recognized visually so as not to be confused with other drugs. Another requirement for a trade name is that it should not have bad connotations in cross-cultural settings. For instance, "Nova" proved to be a bad brand name for a Chevrolet car in Spanish-speaking countries ("No va" means "doesn't go" in Spanish). Of course, regulatory clearance is required from USAN, WHO, EMEA, and international copyright organizations.

In the end, many people were glad that Lyrica was chosen as the brand name for pregabalin. Lyrica projects musical, healing, and soothing qualities, and was a better choice than the initially proposed Alond and Brillior. Partially thanks to a good name, in 2010, Lyrica enjoyed annual sales of $3 billion, a genuine blockbuster drug.

Lyrica is now one of the few drugs that have been approved to treat fibromyalgia. In October 2011, Pfizer stopped a clinical trial of its epilepsy medicine Lyrica after results showed it was effective as a standalone therapy for certain types of seizures. Initially, Lyrica was approved as an adjunctive therapy in combination with other drugs against partial seizures, the most common type experienced by people with the disorder. Partial seizures are those limited to one side of the brain. Lyrica has a linear dosage response, whereas Dilantin does not, which makes it difficult to determine the right dose for Dilantin.

Lyrica is approved for use in combination with other treatments for epilepsy and for FMS.

An added benefit of Lyrica is that it helps to improve the quality of sleep. It actually enhances slow-wave sleep, the kind of sleep you get at the early stages. This attribute makes Lyrica a good sleep aid. Many patients treated with Lyrica for pain actually reported that the most meaningful change in their lives was that they were sleeping better and felt more rested. In comparison, Valium actually reduces slow-wave sleep and lowers the quality of sleep.

§6.10. Lyrica, the Blockbuster Windfall

Not only is Lyrica bringing in billions of dollars in sales for Pfizer, it also brought a windfall to Northwestern University and Silverman himself.[17]

In 2007, after Northwestern collected more than $70 million in royalties for the drug, the university sold a portion of its royalty rights for an additional $700 million.[17] Around the same time, Silverman and his family donated a portion of their earnings from the drug to fund construction of a new Northwestern science building. The facility, which opened in 2009, houses chemistry, biology, and engineering research groups devoted to biomedical science. It is called the Silverman Hall for Molecular Therapeutics & Diagnostics.

Prof. Silverman has published more than 250 papers in organic chemistry, medicinal chemistry, and enzymology. He is also the author of three books, including *The Organic Chemistry of Drug Design and Drug Action*, now in its second edition. In 2009, the Society of Chemical Industry awarded Silverman the Perkin Medal, one of the chemical industry's most prestigious awards. In 2011, the American Chemical Society awarded Silverman the E. B. Hershberg Award for Important Discoveries in Medicinally Active Substances, sponsored by Schering-Plough Research Institute.

Silverman and Northwestern are not the only ones who reaped financial windfalls from their drug discoveries. Prof. Dennis C. Liotta and Emory University earned $540 million in royalties from their discovery of emtricitabine, an HIV medication marketed by Gilead Sciences as Emtriva. Bob Vince and the University of Minnesota were given $400 million for their contributions in making Ziagen (abacavir), a nucleotide HIV drug marketed by GlaxoSmithKline. After retiring from Princeton University, Edward C. Taylor earned a large sum for Princeton from his discovery of pemetrexed (Alimta), an antifolate marketed as an antitumor medication by Eli Lilly. Finally, Taxol (paclitaxel), an anticancer agent once manufactured using chemist Robert A. Holton's semisynthesis method, brought some $350 million to Florida State University.

Today, new drugs are harder and harder to come by as the low-hanging fruits have been harvested. Drug discovery is getting more attention from academia, and more professors are conscientiously turning their basic research toward medicine. More important, a closer collaboration between the drug industry and academia will probably be more synergistic. Historically, the drug industry's golden age (the last three decades of the 20th century) was the direct result of successful, fundamental, and basic research, such as understanding diseases on molecular levels and the elucidation of novel molecular targets for enzymes, receptors, nucleic acid, and structural proteins, and so forth. I sincerely hope these kinds of efforts will bring the next wave of new medicines.

§6.11. How Does Lyrica Work, Exactly?

According to Silverman himself:

> ... This would have been an outstanding example of how the study of fundamental science was rationally developed into an important product, except for one unexpected finding: the anticonvulsant activity of Lyrica is not related to its ability to activate GAD![12]

In other words, Silverman's original theory was incorrect with regard to Lyrica.

Parke-Davis began to investigate the mechanism as soon as Neurontin was on the market. The elucidation work was carried out in their sites at Ann Arbor and Cambridge, U.K., in 1990. After Neurontin was on the market, there was still little understanding of how gabapentin worked.

This is what they knew at the time: Gabapentin and pregabalin were inactive in a large panel of in vitro radio-labeled ligand-binding assays. They were not active in most in vitro functional assays. So this is a unique drug. One of the groundbreaking experiments that Parke-Davis carried out was a radio-labeled gabapentin protein-binding purification. Jason Brown, a graduate student at Parke-Davis's Cambridge site, worked with N. S. Gee. Brown went to a slaughterhouse, bought pounds of fresh pork brain, and ground it up. After incubating it with Neurontin, Brown dissolved the membrane and the solubilized membrane proteins in detergent and eluted them through affinity columns. After going through four different columns, he found only one radio-labeled amino acid terminal. He compared it with the database and identified it as the α2δ (alpha-2-delta) protein, a subunit of calcium channels. The α2δ subunit is an auxiliary protein sticking out of a calcium channel. One more shocking revelation was that these α2δ compounds work only if they are also L transporter substrates, which carry the molecules to the brain. Therefore, binding at the α2δ site is required for their analgesic effects.[18,19]

It is not often that a novel mechanism, let alone a major class of drugs that turned out to be very useful as anticonvulsants and analgesics, is discovered by industrial scientists.

7

Reflections

It is tough to make predictions, especially about the future.

YOGI BERRA

§7.1. What Makes a Blockbuster Drug?

Most blockbusters have at least one thing in common—they are widely prescribed to treat a common illness, such as hypertension, high cholesterol, pain, ulcers, allergies, and depression. The larger the potential patient population, the higher the likelihood for a drug to become a blockbuster.

Obviously, the drug has to be effective. If it does not work, or works only marginally, the average patient is not going to be enthusiastic about taking it.

However, efficacy alone is not good enough. For instance, the old tricyclic antidepressants worked if patients finished the full course of treatment. Unfortunately, they were nonselective, hitting many targets. As a consequence, they were so toxic and replete with side effects that only a fraction of patients (less than 20 percent) were able to tolerate them until completion of a treatment course. In contrast, the newer selective serotonin reuptake inhibitors (SSRIs) are more selective and thus possess fewer side effects. As a result, SSRIs like Prozac, Zoloft, Paxil, Wellbutrin, and revolutionized the treatment of depression. Similarly, atypical antipsychotics such as Zyprexa, Geodon, and others Abilify helped schizophrenia patients tremendously.

These drugs transformed debilitating diseases into treatable chronic illnesses. Most psychiatric patients these days do not need to be institutionalized as they were half a century ago. A drug with good efficacy and a high safety profile treating a widespread disease does not necessarily sell itself, however. Direct advertisements to consumers and a strong sales force are essential to create and sustain the popularity of a drug.

§7.2. How Do You Sustain Blockbuster Drugs?

Blockbuster drugs save lives and support a vibrant pharmaceutical industry. What do we, as a society, do to ensure their sustainability?

While realizing that there is no single prescription for this problem, one could certainly begin by talking about patent reform.

The current patent system is antiquated as far as innovative drugs are concerned. Decades ago, 17 years of patent life was somewhat adequate for drug companies to recoup their investments in R&D because the life cycle from discovery to marketing at the time was relatively short and the cost was lower. Today's drug R&D is a completely new ball game. First of all, the low-hanging fruits have been harvested, and it is becoming more and more challenging to bring novel drugs to the market, especially the ones that are "first-in-class" medicines. Second, the bars that the FDA sets to approve drugs are getting higher and higher, especially with regard to safety after the demises of Vioxx and Avandia. As a consequence, clinical trials are longer and use more patients, increasing costs and eating into patent lives. The latest statistics indicate that it takes $1.3 billion to take a drug from idea to market after taking other failed drugs' costs into account. Therefore, today's patent life of 20 years (since 1995) is insufficient for new medicines, especially the ones that are first-in-class.

The patent life for drugs, especially the first-in-class ones, should be extended because, with such drugs, the risk is at its highest and so is the failure rate. For "me-too" drugs, current patent life might be adequate. Since the life cycle from idea to regulatory approval is getting longer and longer, it would make more sense if the patent clock started ticking after the drug is approved, while exclusivity is still provided after the filing.

It is recognized, of course, that some patients, generic drug companies, and insurance companies may prefer an even shorter patent life for innovative drugs. However, that is akin to killing the goose that laid golden eggs—with short-term gain and long-term loss. The lack of resources will slow the emergence of new drugs.

The current compensation system for the discovery of life-saving drugs is in dire need of reform as well.

When Parke-Davis started marketing Benadryl in 1946, its inventor, George Rieveschl, received a 5 percent royalty for 17 years before the patent expired. That made him a very rich man. Let's fast-forward to 1997, when Parke-Davis began selling Lipitor: Its inventor, Bruce Roth, received an insignificant award in comparison to its billions of dollars in sales. The profits from Lipitor, the largest blockbuster drug ever, went mostly to shareholders and company executives. Although it is reasonable for stockholders to gain financially from investing in drug companies because they take a risk with their money, the top executives' compensations these days are not so reasonable.

Take former Pfizer CEO Henry (Hank) McKinnell, for example. In 2005, while the Pfizer share price had fallen 44 percent since he became CEO in 2001, he took more than $275 million for his retirement package and deferred compensation. At

Pfizer's annual general meeting in 2006 in Denver, an airplane flew over the conference center, trailing a banner that read "Give It Back, Hank!"

In contrast, many drug inventors did not fare so well. Frank Villani, the inventor of Chlor-Trimeton, Optimine, and Claritin, was "eased" out the door by Schering-Plough. In 2007, when Pfizer closed its Ann Arbor site, Bruce Roth was laid off right after the CEO was given a golden parachute. The last few years also saw quite a few inventors of major drugs being laid off or asked to "retire" early. The "rationale" was that they were unlikely to discover more blockbuster drugs after their initial successful discoveries, according to some.

Something is wrong with this picture.

§7.3. The Rise and Decline of the Pharmaceutical Industry

The phenomenon of blockbuster drugs heralded the golden age of the pharmaceutical industry. Patients were happy because taking medicine was vastly cheaper than a hospital stay. Shareholders were happy because huge profits were made and stocks for "big pharma" were considered a sure bet. On the other hand, the drug industry expanded and employed increasingly more and more scientists in its workforce. That employment in turn encouraged academia to train more students in science.

The last 10 years have seen dramatic changes in the pharmaceutical industry. Many patents, especially the ones for blockbuster drugs, have expired. Yet new blockbuster drugs are few and far between, certainly not enough to fill the gap of lost revenues due to patent expirations. The industry has panicked, making many knee-jerk decisions with dubious consequences. One is merger mania. Pfizer alone gobbled up Warner-Lambert,

Pharmacia, and Wyeth, leaving thousands of employees scrambling for employment, many outside the drug industry. Merck swallowed Schering-Plough. The mergers also resulted in site closures of many famous drug R&D centers. Parke-Davis in Ann Arbor, Michigan; Bayer in West Haven, Connecticut; and Roche in Nutley, New Jersey, are only a few of the larger and better-known examples.

Another trend we are seeing these days is outsourcing. Initially, some work was farmed out to small and domestic contract research organizations (CROs). The last decade saw an expansion of CROs for pharmaceutical R&D in both China and India. Recently, even China and India seem to be getting more and more expensive, and some companies have begun to look into other, even cheaper, countries.

Wall Street seems to encourage M&As and spin-offs for the troubled drug industry. In early 2013, Abbott separated its pharmaceutical division, creating a new company named Abbvie. The remaining divisions, including diagnostics and generic drugs, kept the Abbott name. Also in early 2013, Pfizer announced that it was spinning off its animal health division with a new company named Zoetis. There are rumors that its generic drug division may be spun off as well.

It seems that the golden age for small-molecule blockbuster drugs is behind us. However, blockbuster drugs for *biologics* are on the rise. Enbrel, Humira, Remicade, Avastin, and others are now all the rage. Most big pharmaceutical companies are in the process of beefing up their biologics franchises while decreasing small-molecule research. Some vowed to have one-third of their revenues coming from biologics, and even others proposed to have half of their products (and revenues) coming from biologics. In addition, because of the difficulties in producing monoclonal antibodies (many biologics are monoclonal antibodies),

FIGURE 7.1 Paul Ehrlich and Emil von Behring. © Deutsche Post.

herculean efforts are being undertaken by big pharma to develop biosimilars.

The debate about small molecules versus biologics is not new. About a century ago, Paul Ehrlich championed small molecules. His friend, then enemy, and then friend again, Emil von Behring, was a staunch supporter of serum therapy. Despite their initial differences over small molecules versus serum therapy, both of them (figure 7.1) won the Nobel Prize: von Behring in 1901 and Ehrlich in 1908.

Now more than a century has gone by. Certainly, the last century was the century of small molecules, comprising more than 90 percent of the blockbuster drugs. Will the 21st century be the century of biologics, and will blockbuster drugs made from biologics cure the woes of the pharmaceutical industry?

I certainly hope so.[1]

APPENDIX

Chemical Structures of the Drugs

Chapter 1 Before the Age of Blockbuster Drugs

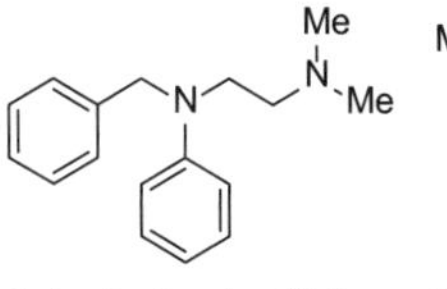

diphenhydramine (Antergan)

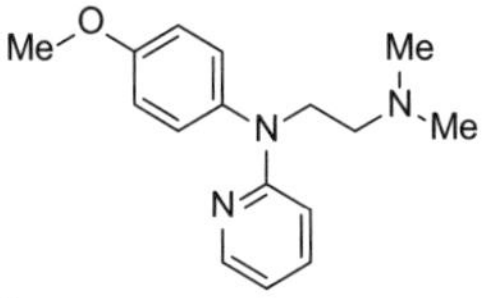

pyrilamine (Neoantergan)

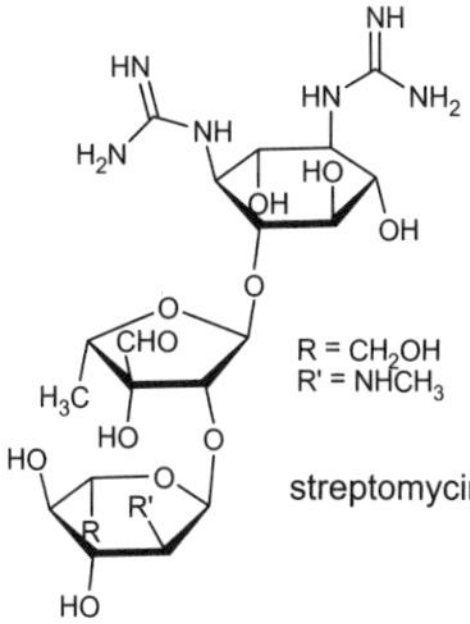

R = CH$_2$OH
R' = NHCH$_3$

streptomycin

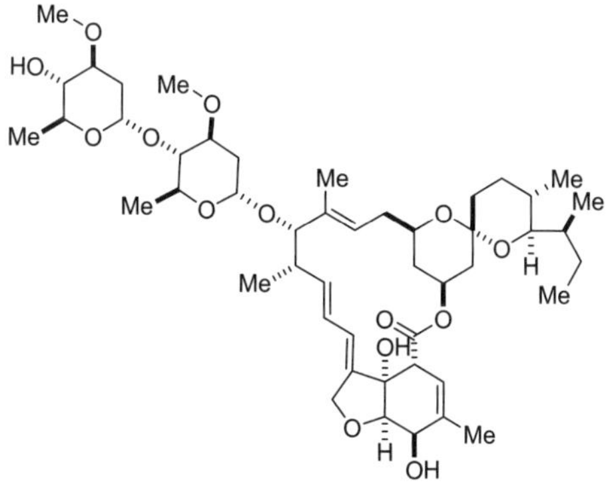

ivermectin (Mectizan)

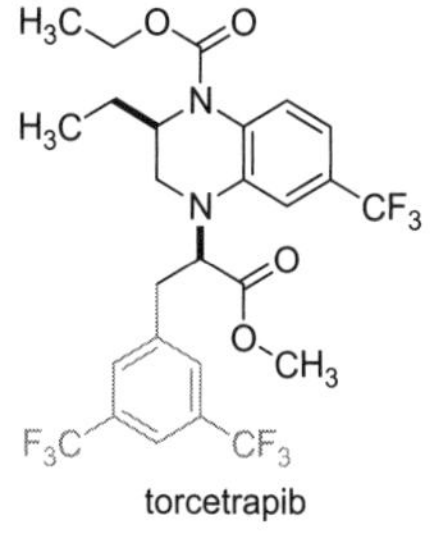

torcetrapib

Chapter 2 Beginning of an Era: The First Blockbuster Drug, Tagamet

amphetamine   chlorpromazine (Thorazine)   adrenaline (epinephrine)   noradrenaline (norepinephrine)

pronethalol   propranolol (Inderal)

practolol (Dalzic)   atenolol (Tenormin)

histidine $\xrightarrow{\text{heat}}$ histamine $+$ $CO_2 \uparrow$

phenbenzamine (Antergan)   pyrilamine (Neoantergan)   4-methyl-histamine   isoprenaline

4-methyl-histamine   isoprenaline   guanylhistamine   burimamide

metiamide   cimetidine (Tagamet)   diazepam (Valium)

albuterol (Ventolin)   labetalol (Trandate)   salmeterol (Serevent)

Chapter 3 More Blockbuster Drugs for Ulcers: Prilosec, Nexium, and Other Proton-Pump Inhibitors

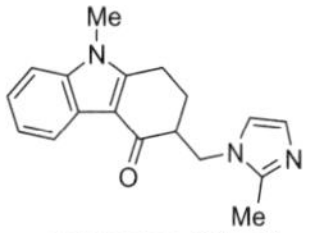

ondansetron (Zofran)

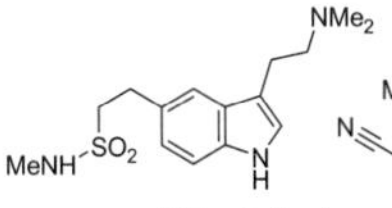

sumastriptan (Imitrex)

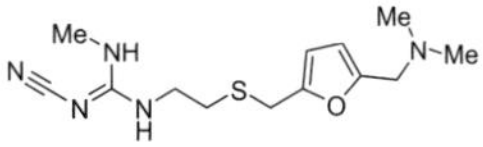

AH18801

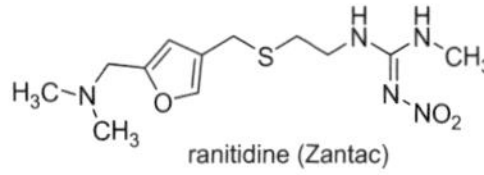

ranitidine (Zantac)

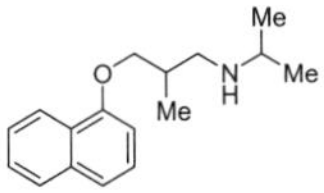

propranolol (Inderal)

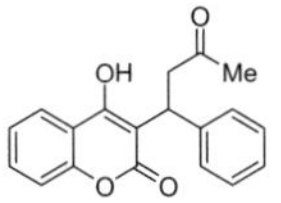

warfarin (Coumadin)

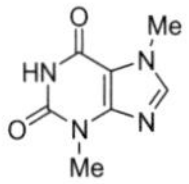

theophylline

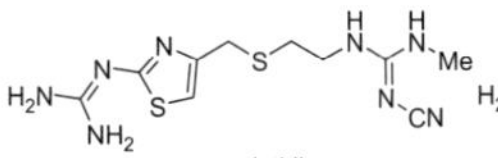

tiotidine

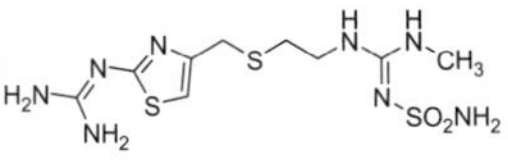

famotidine (Pepcid)

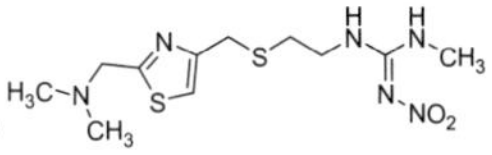

nizatidine (Axid)

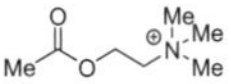

acetylcholine

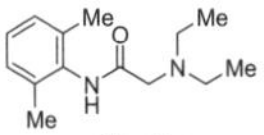

lidocaine

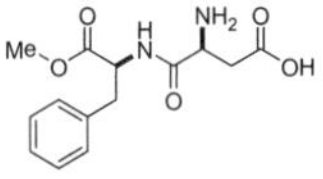

aspartylphenylalanine methyl ester

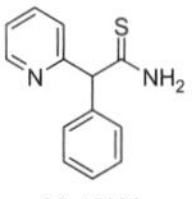

SC-15396

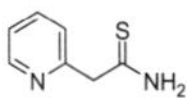

MN131C

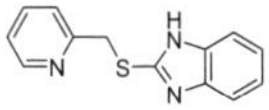

H 124/26

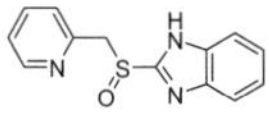

H 83/69 (timoprazole)

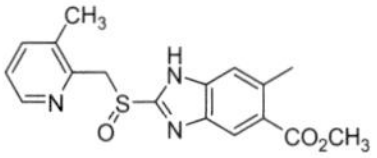

picoprazole

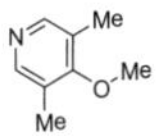

3,5-dimethyl-4-methoxy-pyridine

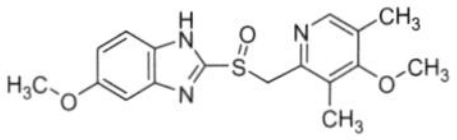

omeprazole (Prilosec)

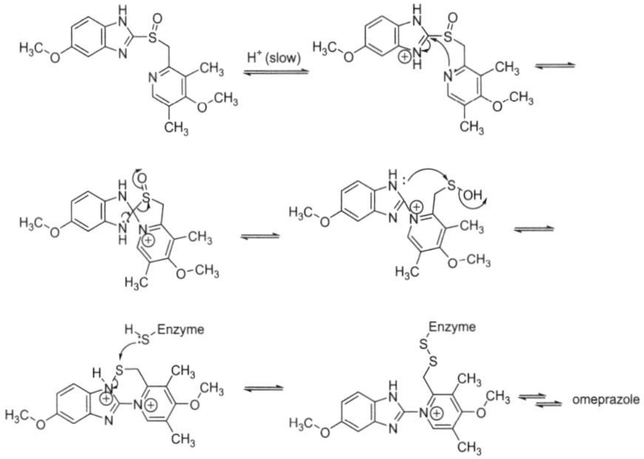

H⁺ (slow)

Enzyme

Enzyme

omeprazole

Bioactivation of omeprazole — the "omeprazole cycle"

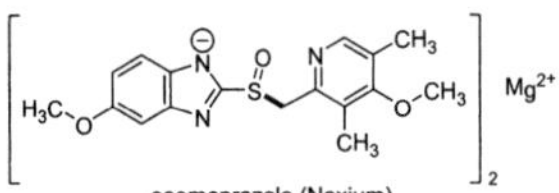

esomeprazole (Nexium)

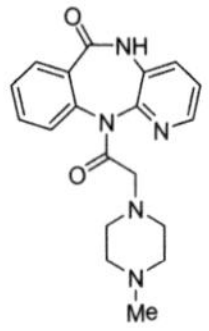

pirenzepine (Gastrozepin)

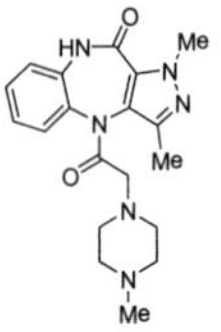

zolenzepine

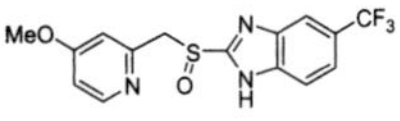

BY319

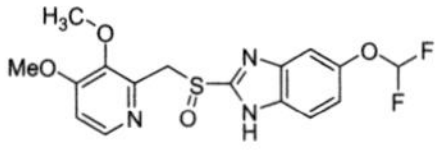

pantoprazole (Protonix)

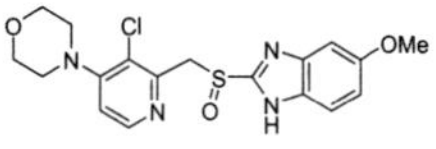

SK&F95601

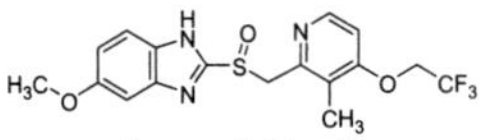

lansoprazole (Prevacid)

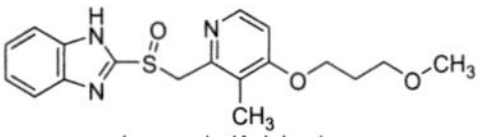

rabeprazole (Aciphex)

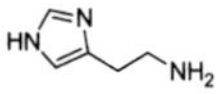

histamine

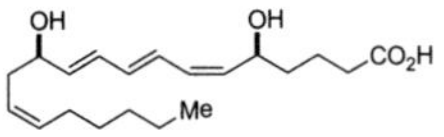

leukotriene B$_4$

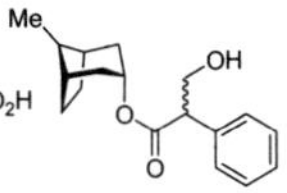

atropine

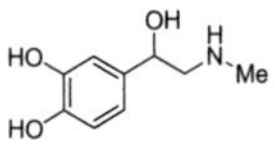

adrenaline (epinephrine)

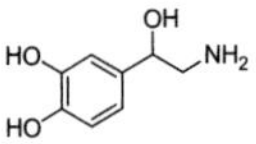

noradrenaline (norepinephrine)

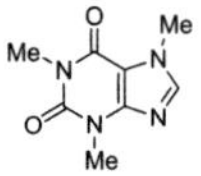

caffeine

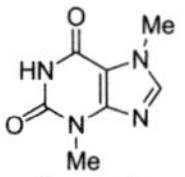

theophylline

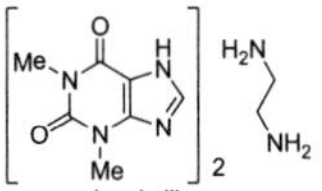

aminophylline

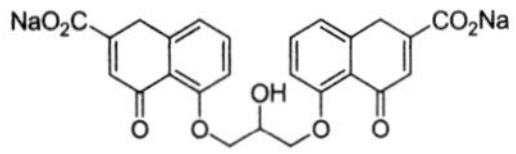

sodium cromoglycate

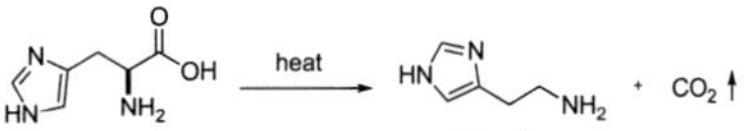

histidine

histamine

Chapter 4 Antihistamines as Allergy Drugs

phenbenzamine (Antergan) pyrilamine (Neoantergan)

diphenhydramine (Benadryl) fluoxetine (Prozac)

meclozine (Dramamine) pheniramine (Trimeton) chlorpheniramine (Chlor-Trimeton) azatadine (Optimine)

terfenadine (Seldane) CYP450 fexofenadine (Allegra)

loratadine (Claritin) desloratadine (Clarinex) cetirizine (Zyrtec)

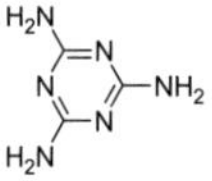

melamine

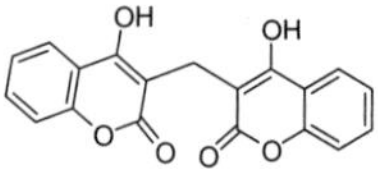

dicumarol

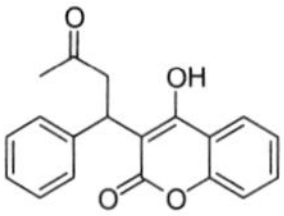

warfarin

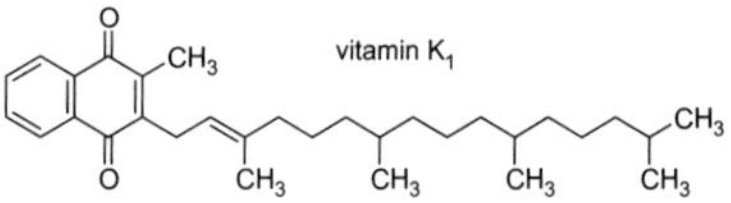

vitamin K₁

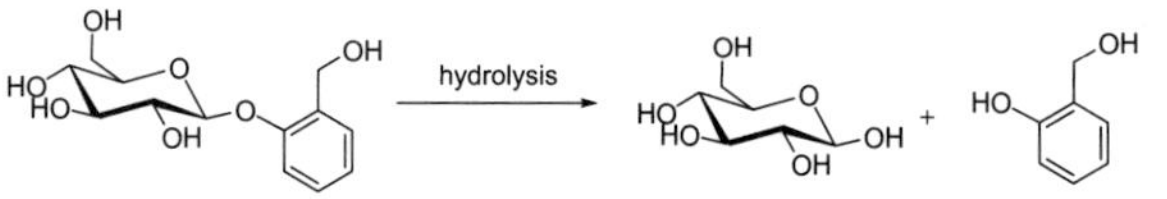

salicin

glucose

salicyl alcohol

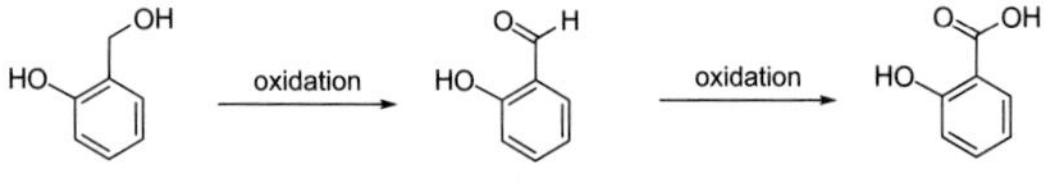

hydrolysis

salicyl alcohol

oxidation

salicyl aldehyde

oxidation

salicylic acid

salicylic acid

acetylation

aspirin (acetylsalicylic acid)

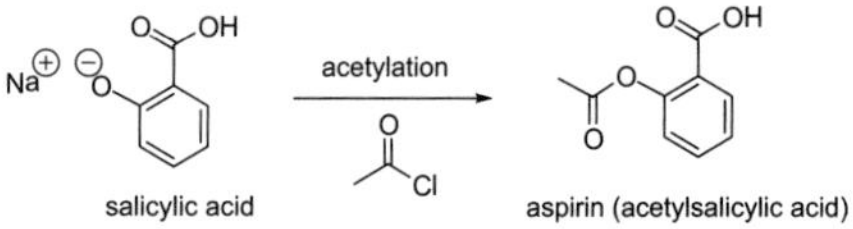

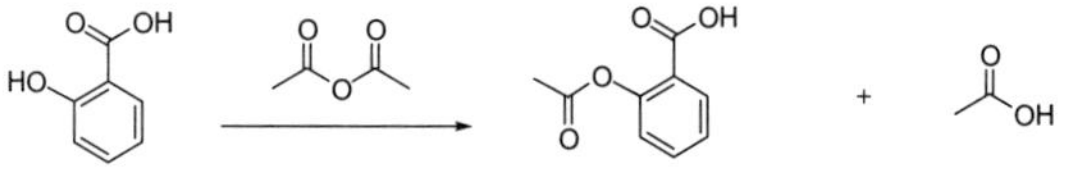

salicylic acid

aspirin (acetylsalicylic acid)

acetic acid

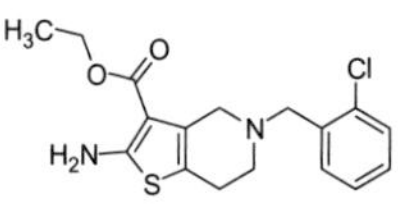

tinoridine (Nonflamin)

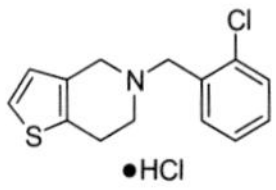

ticlopidine hydrochloride (Ticlid)

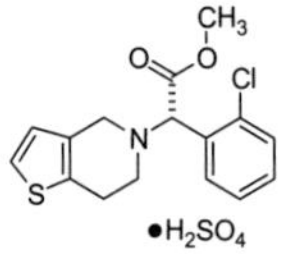

clopidogrel sulfate (Plavix)

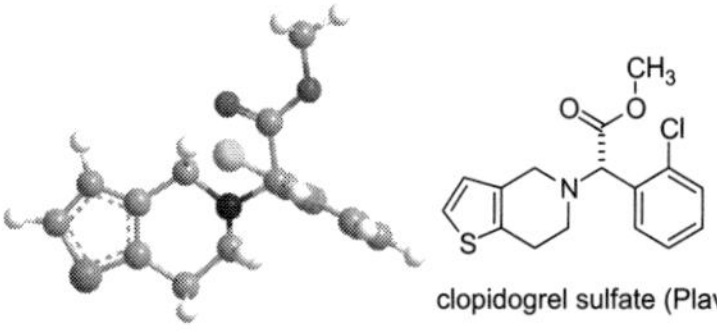

clopidogrel sulfate (Plavix)

Chapter 5 Blood Thinners: From Heparin to Plavix

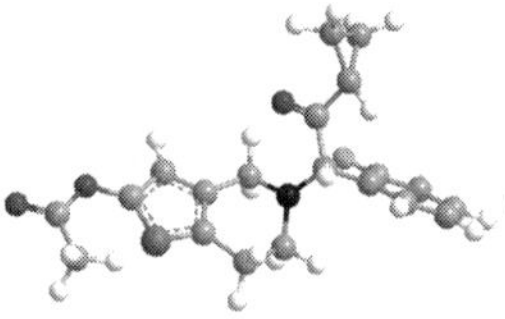

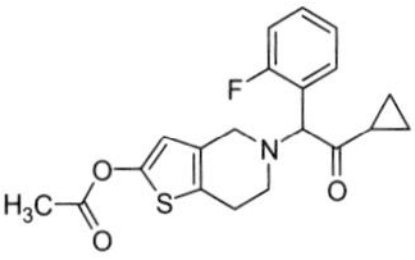

prasugrel (Effient)

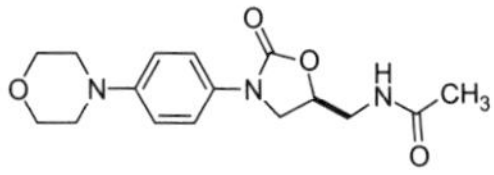

(linezolid) Zyvox

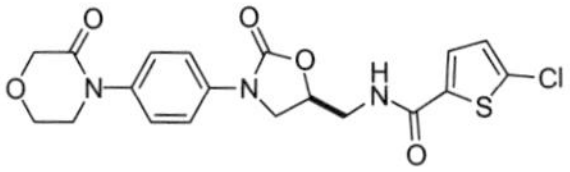

(rivaroxaban) Xarelto

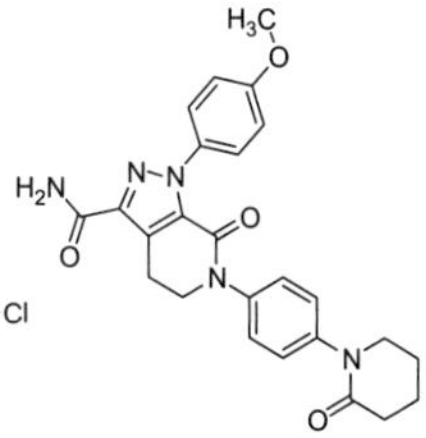

apixaban (Eliquis)

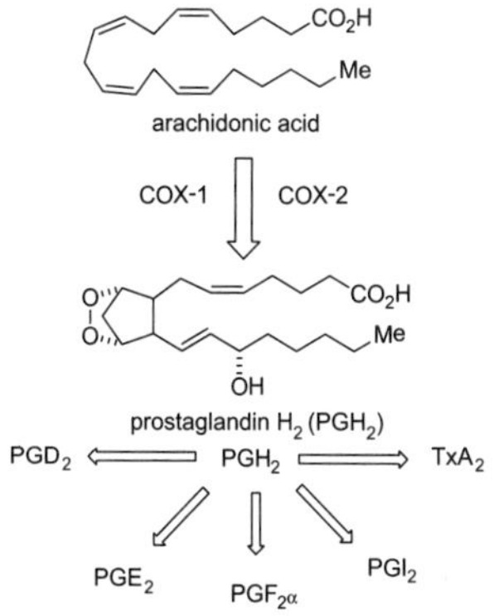

arachidonic acid

COX-1   COX-2

prostaglandin H$_2$ (PGH$_2$)

PGD$_2$ PGH$_2$ TxA$_2$

PGE$_2$ PGF$_{2\alpha}$ PGI$_2$

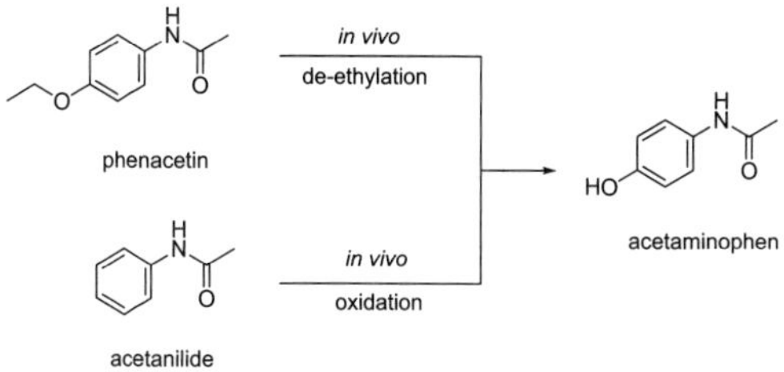

phenacetin

in vivo
de-ethylation

acetanilide

in vivo
oxidation

acetaminophen

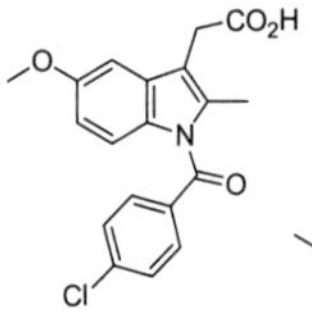

indomethacin

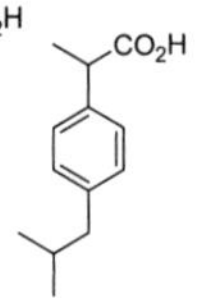

ibuprofen

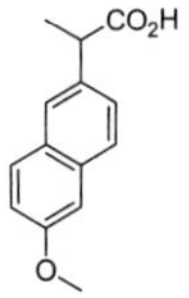

naproxen

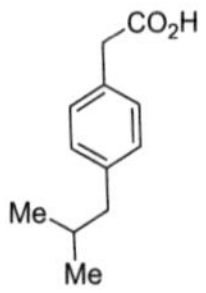

ibufenic

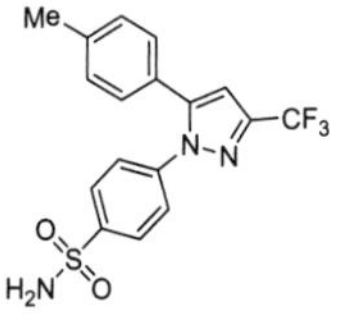

celecoxib (Celebrex)

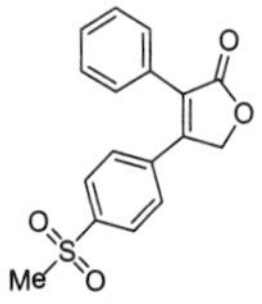

rofecoxib (Vioxx)

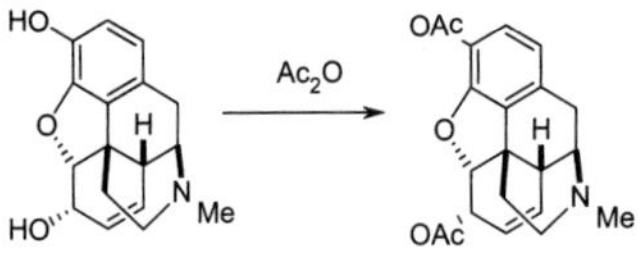

morphine

heroin

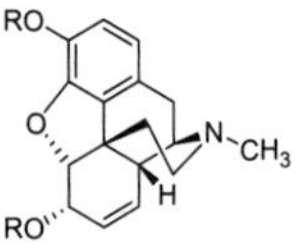

R = H, morphine(Roxanol)
R = COCH₃, heroin

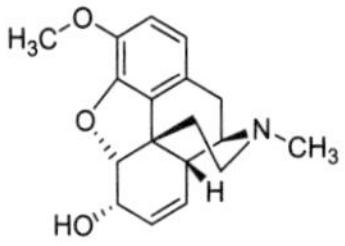

codeine

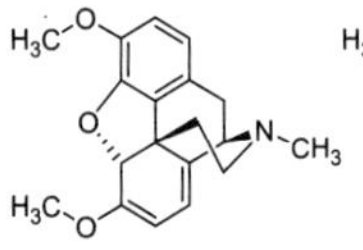

thebaine

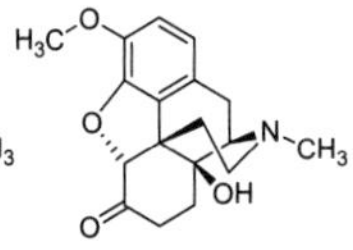

oxycodone (Oxycontin)

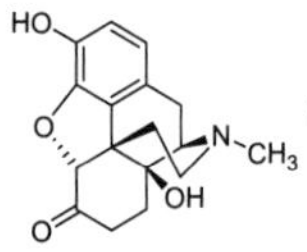

oxymorphone (Opana)

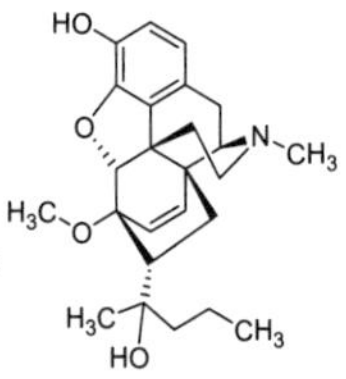

etorphine (Immobilon)

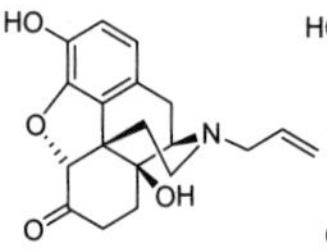

naloxone (Narcan)

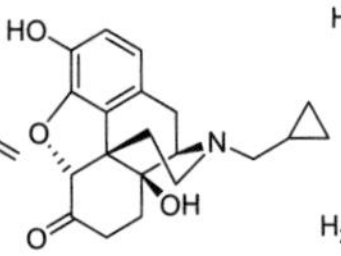

naltrexone (Revia)

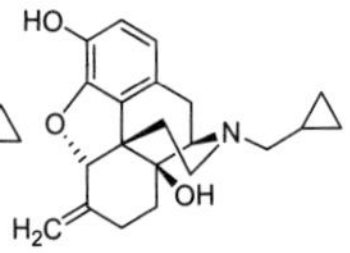

nalmefene (Revex)

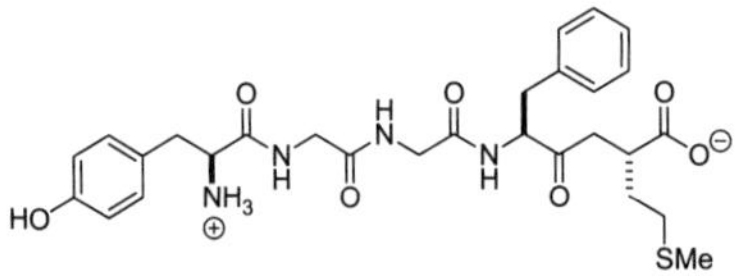

endogenous opioid: met-enkephalin

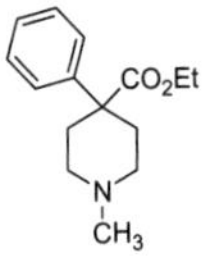

meperidine (Demerol)

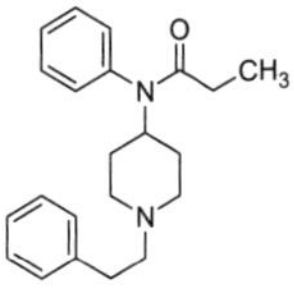

fentanyl (Duragesic)

Chapter 6 Conquest of Pain: Analgesics: From Morphine to Lyrica

sufentanil (Sufenta)   propoxyphene (Darvon)   methadone (Dolophine)   loperamide (Imodium)

Bibliography

Chapter 1 Before the Age of Blockbuster Drugs

1. Dennis Hevesi, "Daniel Bovet, 85, the Discoverer of First Antihistamines, Is Dead," *New York Times*, April 11, 1992.

2. Harry Schwartz, 1989, *Breakthrough: The Discovery of Modern Medicines at Janssen*, Morris Plains, NJ: Skyline Publishing Group, p. 36.

3. Fran Hawthorne, 2003, *The Merck Druggernaut: The Inside Story of a Pharmaceutical Giant*, New York: Wiley, pp. 20–23.

4. The fate of Jeffrey Kindler himself was chronicled in "Inside Pfizer's Palace Coup," *Fortune*, July 28, 2011, p. 13.

5. Arlene Weintraub, 2009, "Will Pfizer's Giveaway Drugs Polish Its Public Image?" *The Business Week*, August 3, 2009, p. 13.

Chapter 2 Beginning of an Era: The First Blockbuster Drug, Tagamet

1. Tom Mahoney, 1959, *The Merchants of Life: An Account of the American Pharmaceutical Industry*, New York: Harper & Brothers, pp. 30–41.

2. Istvan Hargittai and Magdolna Hargittai, 2002, "James W. Black," in *Candid Science II: Conversations with Famous Biomedical Scientists*, London, UK: Imperial College Press.

3. Raymond P. Ahlquist, "Agents Which Block Adrenergic β-Receptors," *Annual Review of Pharmacology* 8, 1968, pp. 259–272.

4. Walter Sneader, 1985, *Drug Discovery: The Evolution of Modern Medicine*, New York: Wiley; Walter Sneader, 2004, *Drug Discovery—A History*, New York: Wiley, p. 159.

5. Dennis Hevesi, "Daniel Bovet, 85, the Discoverer of First Antihistamines, Is Dead," *New York Times*, April 11, 1992.

6. B. Folkow, K. Haeger, and G. Kahlson, "Observations on Reactive Hyperaemia as Related to Histamine, on Drugs Antagonizing Vasodilation Induced by Histamine and on Vasodilator Properties of Adenosine Triphosphate," *Acta Physiol. Scand.* 15, 1948, pp. 264–278.

7. Lawrence Altman, "Ulcer Cases and Surgery Decline; Drug Seems to Control Severity of Illness," *Wall Street Journal*, July 28, 1981.

8. James W. Black, 1988, *Autobiography*, Nobel Foundation.

9. William A. M. Duncan and Michael E. Parsons, "Reminiscences of the Development of Cimetidine," *Gastroenterology* 78, 1980, pp. 620–625; William A. M. Duncan, "Some Decisions in the Development of Cimetidine," *Drug Development Research* 30, 1993, pp. 18–23.

10. Tonse N. K. Raju, "The Nobel Chronicles. 1988: James Whyte Black, (b. 1924), Gertrude Elion (1918–1999), and George H Hitchings (1905–1998)," *Lancet* 355, 2000, p. 1022.

11. Herdis K. M. Molinder, "The Development of Cimetidine: 1964–1976, A Human Story," *Journal of Clinical Gastroenterology* 19, 1994, pp. 248–254.

12. P. Ranganath Nayak and John M. Ketteringham, 1986, *Breakthrough!* New York: Rawson Associates, p. 102.

13. C. Robin Ganellin, 1983, *Cimetidine in Chronicles of Drug Discovery: Volume 1* (ACS Professional Reference Books), J. S. Bindra, and D. Lednicer, eds., New York: Wiley, pp. 1–38.

14. Nina Hall, "A Landmark Drug Design," *Chemistry in Britain*, December 25–27, 1997.

15. Michael Freemantle, "Tagamet, in The Top Pharmaceuticals That Changed the World," *Chemical & Engineering News* 83, 2005, p. 25.

16. P. Ranganath Nayak and John M. Ketteringham, 1986, *Breakthrough!* New York: Rawson Associates, p. 116.

17. Ibid., p. 121.

18. C. Robin Ganellin, 1994, "Discovery of Cimetidine, Ranitidine and Other H2-Receptor Histamine Antagonists," in *Medicinal Chemistry: The Role of Organic Chemistry in Drug Research*, C. R. Ganellin and S. M. Roberts, eds., London, UK: Academic Press, pp. 228–254.

19. "Tagamet": A Medicine That Changed People's Lives," *Chemical & Engineering News*, 2004.

20. A. I. Morris, "The Success of Histamine-2 Receptor Antagonists," *Scandinavian Journal of Gastroenterology* 27(S194), 1992, pp. 71–75.

21. Harvey V. Fineberg and Laurie A. Pearlman, "Surgical Treatment of Peptic Ulcer in the United States: Trends Before and After the Introduction of Cimetidine," *Lancet* 1(8233), 1981, pp. 1305–1307.

22. C. R. Ganellin, 2006, "Discovery of the Antiulcer Drug Tagamet," in *Drug Discovery and Development, Volume 1: Drug Discovery*, M. S. Chorghade, ed., New York: Wiley, pp. 295–311.

23. D. N. R. Kleinfiel, "SmithKline: One-Drug Image," *Wall Street Journal*, May 29, 1984.

24. G. Tweedale, 1990, *At The Sign of the Plough: Allen & Hanburys and the British Pharmaceutical Industry, 1915–1990*, London: John Murray, p. 206.

25. John Bradshaw, 1993, *Ranitidine in Chronicles of Drug Discovery: Volume 3* (ACS Professional Reference Books), D. Lednicer, ed., New York, Wiley, pp. 45–81.

26. Roy T. Brittain, "Discovery and Evolution of Ranitidine," *Current Clinical Practice Series* 1, 1982, pp. 5–15.

27. Takuji Hara, 2003, *Innovation in the Pharmaceutical Industry: The Process of Drug Discovery and Development*, Gloster, UK: Edward Elgar, pp. 94–99.

28. C. R. Ganellin, 2006, "Development of Anti-Ulcer H2-Receptor Histamine Antagonists," in *Analogue-Based Drug Discovery*, J. Fischer and C. R. Ganellin, eds., Weinheim, DE: Wiley-VCH, pp. 71–80; C. R. Ganellin, "Robin Ganellin Gives His Views on Medicinal Chemistry and Drug Discovery," *Drug Discovery Today* 9, 2004, pp. 158–160.

29. Robert P. Bauman, Peter Jackson, and Joanne T. Lawrence, 1997, *From Promise to Performance–A Journey of Transformation at SmithKline Beecham*, Boston: Harvard Business School Press.

Chapter 3   More Blockbuster Drugs for Ulcers: Prilosec, Nexium, and Other Proton-Pump Inhibitors

1. Gina Maranto, "As Acid Reflux Cases Rise, Doctors Are Asking Why," *Wall Street Journal*, December 11, 2001.

2. Leon Jaroff, "Different Ways to Spell Relief," *Wall Street Journal*, November 6, 1995.

3. Irvin M. Modlin, "George Sachs—'I Did It My Way,'" *Journal of Clinical Gastroenterology* 40, 2006, pp. 867–869.

4. Henry M. Sarau, James Foley, George Moonsammy, Virgil D. Wiebelhaus, and George Sachs, "Metabolism of Dog Gastric Mucosa. I. Nucleotide Levels in Parietal Cells," *Journal of Biological Chemistry* 250, 1975, pp. 8321–8399.

5. Allen L. Ganser and John G. Forte, "Potassium Ion–Stimulated ATPase in Purified Microsomes of Bullfrog Oxyntic Cells," *Biochimica et Biophysica Acta, Biomembranes* 307, 1973, pp. 169–180.

6. Barry A. Berkowitz and George Sachs, "Life Cycle of a Blockbuster Drug: Discovery and Development of Omeprazole (Prilosec)," *Molecular Interventions* 2, 2002, pp. 6–11.

7. Sven Erik Sjöstrand, Lars Olbe, and Erik Fellenius, 1999, "The Discovery and Development of the Proton Pump Inhibitors," in *Proton Pump Inhibitors (Milestones in Drug Therapy)*, Lars Olbe, ed., Basel: Birkhäuser Verlag, pp. 3–20.

8. Enar Carlsson, Per Lindberg, and Sverker von Unge, "Two of a Kind," *Chemistry in Britain* 38, 2002, pp. 42–45.

9. Lars Olbe, Enar Carlsson, and Per Lindberg, "A Proton-Pump Inhibitor Expedition: The Case Histories of Omeprazole and Esomeprazol," *Nature Reviews Drug Discovery* 2, 2003, pp. 132–139.

10. Irvin M. Modlin and George Sachs, 2000, *The Logic of Omeprazole: Treatment by Design*, Philadelphia: CoMed Communications.

11. Robert H. Mazur, James M. Schlatter, and Arthur H. Goldkamp, "Structure-Taste Relationships of Some Dipeptides," *Journal of the American Chemical Society* 91, 1969, pp. 2684–2691.

12. Edwin Rabon, Hsuan Hung Chang, Gaetano Saccomani, and George Sachs, "Transport Parameters of Gastric Vesicles," *Acta Physiologica Scandinavica, Supplementum (Proc. Symp. Gastric Ion Transp., 1977)*, 1978, pp. 409–426.

13. Per Lindberg, Arne Brändström, Björn Wallmark, Hillevi Mattsson, Leif Rikner, and Kurt Jurgen Hoffmann, "Omeprazole: The First Proton Pump Inhibitor," *Medicinal Research Reviews* 10, 1990, pp. 1–54.

14. Per Lindberg and Enar Carlsson, 2006, "Esomeprazole in the Framework of Proton-Pump Inhibitor Development," in *Analogue-Based Drug Discovery*, J. Fischer and C. R. Ganellin, eds., Weinheim, DE: Wiley-VCH, pp. 81–113.

15. "AP: Experimental Ulcer Drug Held Superior to Current Mainstay," *New York Times*, January 13, 1989.

16. P. Lindberg, 2006, "Omeprazole," in *Comprehensive Medicinal Chemistry II*, John B. Taylor and David J. Triggle, eds., Oxford, UK: Elsevier, pp. 213–225.

17. Gardiner Harris, "2 New Fronts in Heartburn Market Battle," *New York Times*, August 20, 2003.

18. Rene J. Dubos, 1960, *Pasteur and Modern Science* (Science Study Series, S15), first edition, Prescott, AZ: Anchor Books.

19. Jörg Senn-Bilfinger and Ernst Sturm, 2006, "The Discovery of a New Proton Pump Inihibitor: The Case History of Pantoprazole," in *Analogue-Based Drug Discovery*, J. Fischer and C. R. Ganellin, eds., Weinheim, DE: Wiley-VCH, pp. 114–121.

20. R. J. Playford, T. Podas, and I. Modlin, "Pantoprazole, Prout and the Proton Pump," *Hospital Medicine* 60, 1999, pp. 500–504.

21. Morton A. Meyers, 2007, *Happy Accidents, Serendipity in Modern Medical Breakthroughs*, New York: Arcade, pp. 99–113.

Chapter 4   Antihistamines as Allergy Drugs

1. G. Mitman, 2008, *Breathing Space: How Allergies Shape Our Lives and Landscapes*, New Haven, CT: Yale University Press.

2. M. Jackson, 2007, *Allergy: The History of a Modern Malady*, Chicago: Reaktion.

3. Margot Slade and Eva Hoffman, "Treating Allergies from the Inside Out," *Wall Street Journal*, February 15, 1981.

4. Tom Yulsman, "Taking Care: Allergies," *Wall Street Journal*, April 17, 1988.

5. M. B. Emanuel, "Histamine and the Antiallergic Antihistamines: A History of Their Discoveries," *Clinical and Experimental Allergy* 29 (Suppl. 3), 1999, pp. 1–11.

6. Dennis Hevesi, "George Rieveschl; Invented Benadryl," *New York Times*, September 30, 2007.

7. James Black, "A Personal View of Pharmacology," *Annual Review of Pharmacology and Toxicology* 36, 1996, pp. 1–33.

8. Bryan B. Molloy, David T. Wong, and Ray W. Fuller, "The Discovery of Fluoxetine," *Pharmaceutical News* 1(2), 1994, pp. 6–10.

9. Tom Mahoney, 1959, *The Merchants of Life: An Account of the American Pharmaceutical Industry*, New York, Harper & Brothers, pp. 253–266.

10. Stephen C. Stinson, "Albert Carr Wins Perkin Medal," *Chemical & Engineering News* 77, 1999, pp. 63–64.

11. Allen Barnett and Michael J. Green, 1993, "Loratadine," in *Chronicles of Drug Discovery, Volume 3*, Daniel Lednicer, ed., Washington, DC: American Chemical Society, pp. 83–99.

12. Stephen S. Hall, "The Claritin Effect; Prescription for Profit," *New York Times*, March 11, 2001.

Chapter 5   Blood Thinners: From Heparin to Plavix

1. W. B. Fye, "Heparin: The Contributions of William Henry Howell," *Circulation* 69, 1984, pp. 198–203.

2. Jay McLean, "The Discovery of Heparin," *Circulation* 19, 1959, pp. 75–78.

3. Jay McLean, "The Thromboplastic Action of Cephalin," *American Journal of Physiology* 41, 1916, pp. 250–257; Jay McLean, "The Relation between the Thromboplastic Action of Cephalin and Its Degree of Unsaturation," *American Journal of Physiology* 43, 1917, pp. 586–596.

4. J. A. Marcum, "The Origin of the Dispute over the Discovery of Heparin," *Journal of the History of Medicine and Allied Sciences* 55, 2000, pp. 37–66.

5. "Jay McClean (1890–1957), Discoverer of Heparin," *Journal of American Medical Association (JAMA)* 201, 1967, p. 144.

6. Charles H. Best, "Preparation of Heparin and Its Use in the First Clinical Cases," *Circulation* 19, 1959, pp. 79–86.

7. Walter Sneader, 1985, *Drug Discovery: The Evolution of Modern Medicines*, Chichester, UK: Wiley, p. 161.

8. Walter Sneader, 2005, *Drug Discovery: A History*, Chichester, UK: Wiley, p. 269.

9. R. J. Baird, "Give us the Tools...:" The Story of Heparin—as Told by Sketches from the Lives of William Howell, Jay McLean, Charles Best, and Gordon Murray," *Journal of Vascular Surgery* 11, 1990, pp. 4–18.

10. J. E. Jorpes, "Heparin: A Mucopolysaccharide and an Active Antithrombotic Drug," *Circulation* 19, 1959, pp. 87–91.

11. M. K. Davies and A. Hollman, "Heparin," *Heart (British Cardiac Society)* 80(2), 1998, p. 120.

12. Bill Power, "Heparin's Deadly Side Effects," *New York Times*, November 13, 2008.

13. Andrew Jacobs, "Chinese Release Increased Numbers in Tainted Milk Scandal," *New York Times*, December 3, 2008.

14. Paul Griminger, "Vitamin K Antagonists: The First 50 Years," *Journal of Nutrition* 117, 1986, pp. 1325–1329.

15. Mike Scully, "Warfarin Therapy: Rat Poison and the Prevention of Thrombosis," *Biochemist* 24, 2002, pp. 15–17.

16. Karl Paul Link, "The Discovery of Dicumarol and Its Sequels," *Circulation* 19, 1959, pp. 97–107.

17. Douglas Wardrop and David Keeling, "The Story of the Discovery of Heparin and Warfarin," *British Journal of Haematology* 141, 2008, pp. 757–763.

18. B. M. Duxbury and L. Poller, "The Oral Anticoagulant Saga: Past, Present, and Future," *Clinical and Applied Thrombosis/Hemostasis: Official Journal of the International Academy of Clinical and Applied Thrombosis/Hemostasis* 7, 2001, pp. 269–275.

19. Jerold A. Last, "The Missing Link: The Story of Karl Paul Link," *Toxicological Sciences: An Official Journal of the Society of Toxicology* 66, 2002, pp. 4–6.

20. Nicole Kresge, Robert D. Simoni, and Robert L. Hill, "Hemorrhagic Sweet Clover Disease, Dicumarol, and Warfarin: The Work of Karl Paul Link," *Journal of Biological Chemistry* 280(8), 2005, pp. e5–e6.

21. W. F. Barker, E. B. Hickman, J. A. Harper, and J. Lungren, "Venous Interruption for Pulmonary Embolism: The Illustrative Case of Richard M. Nixon," *Annals of Vascular Surgery* 11, 1997, pp. 387–390.

22. M. Pirmohamed, "Warfarin: Almost 60 Years Old and Still Causing Problems," *British Journal of Clinical Pharmacology* 62, 2006, pp. 509–511.

23. Pollack, "Andrew Gene Test for Dosage of Warfarin Is Rebuffed," *New York Times*, May 5, 2009.

24. E. G. L. Bywaters, 1963, "The History of Salicylates," in *Salicylates, An International Symposium*, A. St. J. Dixon, B. K. Martin, M. J. H. Smith, and P. H. N. Wood, eds., Boston: Little, Brown, p. 3.

25. Richard L. Mueller and Stephen Scheidt, "History of Drugs for Thrombotic Disease: Discovery, Development, and Directions for the Future," *Circulation* 89, 1994, pp. 432–449.

26. Diarmuid Jeffreys, 2004, *Aspirin: The Remarkable Story of a Wonder Drug*, New York: Bloomsbury.

27. H. J. Weiss, "The Discovery of the Antiplatelet Effect of Aspirin: A Personal Reminiscence," *Journal of Thrombosis and Haemostasis* 1, 2003, pp. 1869–1875.

28. J. R. Vane and R. M. Botting, "The Mechanism of Action of Aspirin," *Thrombosis Research* 110, 2003, pp. 255–258.

29. Jerry L. Bauman, "Tales of Two Oral Anticoagulants from Natural Product Research and Their Impact on Clinical Pharmacy," *Pharmacotherapy* 24(10 Pt. 2), 2004, pp. 166S–168S.

30. David Gustafsson, Ruth Bylund, Thomas Antonsson, Ingemar Nilsson, Jan-Erik Nystroem, Ulf Eriksson, Ulf Bredberg, and Ann-Catrine Teger-Nilsson, "Case History: A New Oral Anticoagulant: The 50-Year Challenge," *Nature Reviews Drug Discovery* 3, 2004, pp. 649–659.

31. David Gustafsson, "Discovery of Ximelagatran in a Historical Perspective," *Seminars in Vascular Medicine* 5, 2005, pp. 227–234.

32. Alex C. Spyropoulos, "Brave New World: The Current and Future Use of Novel Anticoagulants," *Thrombosis Research* 123(Suppl. 1), 2008, pp. S29–S35.

33. J. P. Maffrand and F. Eloy, "Synthesis of Thienopyridines and Furopyridines of Therapeutic Interest," *European Journal of Medicinal Chemistry* 9, 1974, pp. 483–486.

34. Steven R. Steinhubl, Walter A. Tan, JoAnne M. Foody, and Eric J. Topol, "Incidence and Clinical Course of Thrombotic Thrombocytopenic Purpura Due to Ticlopidine Following Coronary Stenting," *Journal of the American Medical Association (JAMA)* 281, 1999, pp. 806–810.

35. Lawrence K. Altman, "Clot Blocker Is Linked to Disorder of the Blood," *New York Times*, April 21, 2000.

36. Sidney H. Stein, U.S.D.J., U.S. District Court Southern District of New York Opinion and Order, 02 Civ. 2255 (SHS), *Sanofi-Synthelabo; Sanofi-Synthelabo, Inc.; and Bristol-Myers Squibb Sanofi Pharmaceuticals Holding Partnership, Plaintiffs, v. Apotex Inc. and Apotex Corp.*, June 19, 2007, New York, NY.

37. *Pharmaceutical Salts: Properties, Selection, and Use*, 2008, P. H. Stahl and C. G. Wermuth, eds., Weinheim, DE: Wiley-VCH.

38. Milt Freudenheim, "Company News; Bristol-Myers Squibb in Alliance to Develop Heart Drugs," *New York Times*, June 2, 1993.

39. CAPRIE Steering Committee (M. Gent, et al.), "A Randomized, Blinded Trial of Clopidogrel versus Aspirin in Patients at Risk of Ischaemic Event," *Lancet* 348, 1996, pp. 1329–1339.

40. David J. Morrow, "International Business; French Drug Makers to Combine in $10.4 Billion Stock Deal," *New York Times*, December 3, 1998.

41. Lawrence Altman, "Drug Hailed as a Heart and Stroke Protector," *New York Times*, March 20, 2001.

42. Bethany McLean, "The High Price of Drug Patents: Apotex's CEO Is Attacking Drug Patent Settlements That He Says Benefit Everyone Except the Consumer," *Fortune*, January 31, 2007.

43. Miriam Schuchman, 2006, *The Drug Trial, Nancy Olivieri and the Science Scandal That Rocked the Hospital for Sick Children*, Toronto: Random House Canada.

44. Edward Wawrzynczak, "Plavix Franchise in Jeopardy," *Nature Reviews Drug Discovery* 5, 2006, pp. 712–713.

45. Miriam Shuchman, "Delaying Generic Competition—Corporate Payoffs and the Future of Plavix," *New England Journal of Medicine* 355, 2006, pp. 1297–1300.

46. Komal Shah Bhukhanwala and R. Godbole, "Clopidogrel Bisulfate (Plavix Sanofi and Bristol-Myers Squibb): Another Billion Dollar Drug under Attack by the Generics," *Expert Opinion on Therapeutic Patents* 16, 2006, pp. 1609–1611.

47. Stephanie Saul, "Marketers of Plavix Outfoxed on a Deal," *New York Times*, August 9, 2006.

48. Pryor Cashman Sherman and Caesar, Rivise, Berstein Cohen & Pokotilow, Ltd. Apotex's Memorandum of Law in Opposition to Plaintiffs' Motion for Preliminary Injunction, in the U.S. District Court for the Southern District of New York, Civil Action No. 02-CV-2255 (RWS) and Civil Action No. 02-CV-3672 (RWS), *Sanofi-Synthelabo; Sanofi-Synthelabo, Inc.; and Bristol-Myers Squibb Sanofi Pharmaceuticals Holding Partnership, Plaintiffs,v. Apotex Inc. and Apotex Corp.*, August 7, 2006, New York, NY.

49. "Judge's Sentence for Former Bristol-Myers Exec: Write a Book," *Wall Street Journal*, June 9, 2009.

50. Matthew Herper, "Blood Thinner's Approval a Muted Win for Lilly," *Fortune.com*, July 10, 2009.

51. Val Brickate Kennedy and Steve Goldstein, "AstraZeneca Shares Jump on Blood-Thinner Drug Study," *Wall Street Journal*, May 11, 2009.

52. Andrew Pollack, "Novartis Buys Rights to a Drug to Thin Blood," *New York Times*, February 12, 2009.

53. "Merck Acquires Rights to Anticoagulant," *Wall Street Journal*, July 9, 2009.

54. Harry R. Buller, Ander T. Cohen, Bruce Davidson, Herve Decousus, Alex S. Gallus, Michael Gent, Gerard Pillion, Franco Piovella, Martin H. Prins, and Gary E. Raskob, "Idraparinux versus Standard Therapy for Venous Thromboembolic Disease," *New England Journal of Medicine* 357, 2007, pp. 1094–1104.

55. Susanne Roehrig, Alexander Straub, Jens Pohlmann, Thomas Lampe, Josef Pernerstorfer, Karl-Heinz Schlemmer, Peter Reinemer, and Elisabeth Perzborn, "Discovery of the Novel Antithrombotic Agent 5-Chloro-N-({(5S)-2-oxo-3- [4-(3-oxomorpholin-4-yl)phenyl]-1,3-oxazolidin-5-yl} methyl)thiophene-2-carboxamide (BAY 59-7939): An Oral, Direct Factor Xa Inhibitor," *Journal of Medicinal Chemistry* 48, 2005, pp. 5900–5908.

56. "FDA Isn't Ready to Approve J&J Anticlotting Drug," *Wall Street Journal*, May 28, 2009.

57. Donald J. P. Pinto, Michael J. Orwat, Stephanie Koch, Karen A. Rossi, Richard S. Alexander, Angela Smallwood, Pancras C. Wong, Alan R. Rendina, Joseph M. Luettgen, Robert M. Knabb, Kan He, Baomin Xin, Ruth R. Wexler, and Patrick Y. S. Lam, "Discovery of 1-(4-Methoxyphenyl)-7-oxo-6-[4-(2-oxo-1-piperidinyl]phenyl]-4,5,6,7-tetrahydro-1H-pyrazolo[3,4-c]pyridine-3-carboxamide (Apixaban, BMS-562247), a Highly Potent, Selective, Efficacious, and Orally Bioavailable Inhibitor of Blood Coagulation Factor Xa," *Journal of Medicinal Chemistry* 50, 2007, pp. 5339–5356.

Chapter 6   Conquest of Pain: Analgesics: From Morphine to Lyrica

1. "NIH Consensus Conference, Acupuncture," *Journal of the American Medical Association (JAMA)* 280, 1998, pp. 1518–1524.

2. J. E. Brody, "Many Treatments Can Ease Chronic Pain," *Wall Street Journal*, November 20, 2007.

3. Martin Booth, 1999, *Opium: A History*, New York: St. Martin's Griffin.

4. John Lowe III, 2013, *CNS Drugs in Drug Discovery: Practices, Processes, and Perspectives*, J. J. Li and E. J. Corey, eds., Hoboken, NJ: Wiley, pp. 245–286.

5. K. C. Nicolaou and T. Montagnon, 2008, *Molecules That Changed the World*, Weinheim, DE: Wiley-VCH, pp. 76–77.

6. Harry Schwartz, 1989, *Breakthrough: The Discovery of Modern Medicines at Janssen*, Morris Plains, NJ: Skyline Publishing Group.

7. Don L. Goldenberg, 1996, *Chronic Illness and Uncertainty: A Personal and Professional Guide to Poorly Understood Syndromes, What We Know and Don't Know about Fibromyalgia, Migraine, Depression and Related Illnesses*, Newton Lower Falls, MA: Dorset Press.

8. C. W. Lindsley, "Molecule of the Month," *Curr. Top. Med. Chem.* 8, 2008, p. 1100.

9. L. P. Rowland, 2008, *The Legacy of Tracy J. Putnam and H. Houston Merritt: Modern Neurology in the United States*, New York: Oxford University Press.

10. G. Satzinger, "Antiepileptics from Gamma-Aminobutyric Acid," *Arzneimittel-Forschung* 44, 1994, pp. 261–266.

11. G. Satzinger, "Drug Discovery and Commercial Exploitation," *Drug News Perspect.* 14, 2001, pp. 197–207.

12. R. B. Silverman, "From Basic Science to Blockbuster Drug: The Discovery of Lyrica," *Angew. Chem. Intl. Ed.* 47, 2008, pp. 3500–3504.

13. R. Andruszkiewicz and R. B. Silverman, "A Convenient Synthesis of 3-Alkyl-4-Aminobutanoic Acids," *Synthesis*, 1989, pp. 953–955.

14. R. Andruszkiewicz and R. B. Silverman, "4-Amino-3-Alkylbutanoic Acids as Substrates for γ -Aminobutyric Acid Aminotransferase," *Journal of Biological Chemistry* 265, 1990, pp. 22288–22291.

15. R. Andruszkiewicz, A. G. M. Barrett, and R. B. Silverman, "Chemo-enzymatic Synthesis of (R)- And (S)-4-Amino-3-Methylbutanoic Acid," *Synthetic Communications* 20, 1990, pp. 159–166.

16. Pat Wechsler, "Pfizer Stops Epilepsy Drug Trial as Efficacy Proved," *Wall Street Journal*, October 7, 2011.

17. Mitch Jacoby, "Royalty Payouts from University-Held Patents Have Power to Transform Chemistry Departments," *C&EN News* 86, 2008, pp. 56–61.

18. Charles P. Taylor, "Mechanisms of Analgesia by Gabapentin and Pregabalin—Calcium Channel α2-δ [Cavα2-δ] Ligands," *Pain* 142, 2009, pp. 13–16.

19. Shelley L. Davies, Óscar Villacañas, and Jordi Bozzo, "Targeting the α2-δ Calcium Channel Subunit for Pain Therapeutics," *Drugs of the Future* 31, 2006, p. 837.

Chapter 7   Reflections

1. The year 2012 was a banner year for both the FDA and the pharmaceutical industry, with 39 new molecular entities approved. The last time the FDA approved more than 39 NDAs was 1996. Twenty-six of the drugs, two-thirds, are small-molecule drugs. Among the 13 biologics, most of them were peptides or enzyme replacements. Only two of them were monoclonal antibodies: Affymax's reginesatide (Omontys), an erythropoietin analogue, for the treatment of anemia related to chronic

kidney disease (erythropoietin, also known as EPO, was the drug Lance Armstrong used for doping); and Genentech's pertuzumab (Perjeta), an HER2 inhibitor for the treatment of HER2-positive breast cancer. Perjeta is a me-too drug of Genentech's trastuzumab (Herceptin). Interestingly, Pfizer alone had five drugs added to its drug portfolio in 2012.

Index

Pliny, 105
Plough Inc., 78. *See also* Schering-
    Plough Corporation
Polysaccharide, 95
Pool, Michael, 163
Popielski, Leon, 13
Poplar, 138
Poppies, 143, *143*, 145, 147
Portola Pharmaceuticals, 132
Post-blockbuster syndrome, 39
Post-operative pain, 137
Potassium bromide, 154
Potential patient population, 168
Po-Wai Yuen, 162
Practolol (Dalzic), 9
Prasit, P., 141
Prasugrel (Effient), 129–32, *130*
Pravachol (pravastatin), 50, 120
Pravastatin (Pravachol), 50, 120
Prazosin, 73
Pregabalin (Lyrica), 153, *160*,
    161–65, 167. *See also*
    Lyrica (pregabalin)
Pregl, Fritz, 100
Prevacid, 44, 48
Prevacid (lansoprazole), 63
Price, Barry, 30
Prichard, Robert, 124–25
Prilosec (omeprazole), 43–57, 62,
    111
    Approval of, 55
    Bioavailability of, 60
    How it works, 55–57
    MOA of, 55–57
    Nexium and, 57, 60–61
    Patent on, 61
Princeton University, 166
Promethazine (Phenergan), 83
Pronethalol, 9

Propoxyphene (Darvon), 148
Propranolol (Inderal), 9, 33
Prostaglandins, 35, 109, 139–40
Proton, 139
Protonix (pantapazole), 44, 48,
    57, 61, 62–63
Proton-pump inhibitors, 41–66.
    See also specific drugs
    Aciphex, 44
    for intravenous use in critical-
        care patients, 63
    "Me-too," 61–62
    Nexium, 44
    Prevacid, 44
    Prilosec, 43–44
    Protonix, 44
Proton pumps, 44–46, 50
Prototypes, use of for drug dis-
    covery, 37
Prout, William, 41
Prozac (fluoxetine), 125
    Benadryl as prototype for, 76
Pschorr, Robert, 147
Publication, 30, 37–38
Pulmonary embolism, 89
Putnam, Tracy Jackson, 154–56
Pyridine, 31, 52, 62, 77
Pyridine rings, 77, 78
Pyrilamine (Neoantergan), 2, 12,
    73, 74, 78
Pyrrole, 31

Quayle, Dan, 89, 96
Quick, Armand, 108
Quinine, 125

R&D (research and develop-
    ment), 6, 169–70, 172
R951, 150